CHANGING AGRICULTURAL SCENARIO IN NORTH-EAST INDIA

Books Published for NEICSSR, Shillong

- Reorganisation of North East India Since 1947/ *B. Datta Ray* and *S.P. Agrawal (Eds.)*
- Sociological Constraints to Industrial Development in North-East India/ *B.K. Datta Ray* and *Prabin Baishya (Eds.)*
- Population, Poverty and Environment in North East India/ *B. Datta Ray, H.K. Mazhari, P.M. Passah* and *M.C. Pandey (Eds.)*
- Agenda for North East India/ *B. Datta Ray (Ed.)*
- Development Priorities in North East India/ *Bimal J. Deb (Ed.)*
- Linguistic Situation in North East India/ *Mrinal Miri (Ed.)*
- Perspective of Security and Development in North East India/ *S.K. Agnihotri* and *B. Datta Ray (Eds.)*
- Ethnic Issues, Secularism and Conflict Resolution in North East India/ *Bimal J. Deb (Ed.)*

About the Editors

Bimal J. Deb is Professor of Political Science in North Eastern Hill University, Shillong.

B. Datta Ray is a freedom fighter and a social scientist. He is Founder Secretary of North East India Council for Social Science Research. He has to his credit two books, namely, *Administrative History of North East India and Tribal Identity* and *Tensions in North East India.* Also, he has edited 17 books on North East India.

CHANGING AGRICULTURAL SCENARIO IN NORTH-EAST INDIA

Edited by

BIMAL J. DEB

B. DATTA RAY

CONCEPT PUBLISHING COMPANY, NEW DELHI-110059

Cataloging in Publication Data—DK
Courtesy: D.K. Agencies (P) Ltd. <docinfo@dkagencies.com>

Seminar on Changing Agricultural Scenario in North East India (2004: Shillong, India)
Changing agricultural scenario in North-east India / edited by Bimal J. Deb, B. Datta Ray.
p. cm.
Papers presented at the Seminar on Changing Agricultural Scenario in North East India, held at Shillong during 12-13 December 2004.
Includes bibliographical references.
Includes index.
ISBN 8180692647

1. Agriculture—India, Northeastern—Congresses. 2. Agriculture—Economic aspects—India, Northeastern—Congresses. I. Deb, Bimal J. II. Datta Ray, B. (Basudeb), 1925- III. Title.

DDC 630.954 1 22

ISBN 81-8069-264-7

First Published 2006

Published and Printed by
Ashok Kumar Mittal
Concept Publishing Company
A/15-16, Commercial Block, Mohan Garden,
NEW DELHI-110059 (India)

Phones: 25351460, 25351794
Fax: 091-11-25357103
Email: publishing@conceptpub.com

FOREWORD

North East India is endowed with rich natural resources of good soil and climate. But because of some social and economic constraints, and for some biophysical limitation, the land resources are not being used properly for agricultural development.

In all the hills, the land system is governed by and large by traditions and customs. Assam, Tripura and Manipur have codified land laws and manuals relating land ownership.

In the hill areas, the terrain is difficult and fertility of the soil is low. There is the problem of soil erosion because of heavy rains and inputs are meagre and primitive. Nearly 80 per cent of the population of the hill areas are hardly able to raise the minimum of their food requirements.

The problems of agriculture in North East India are manifold. The lack of uniformity of soil condition, the continuance of the practice of shifting cultivation in wide areas, the absence of irrigation facilities, the serious problem of flood affecting large areas of agricultural land and the lack of marketing and credit facilities for the agricultural produce are only few of them. The scope of large scale mechanisation and wide use of electric power for agricultural development in the region is very limited. Agro-industrial development in North East India is a distant dream.

The economy of North East India is dominated by primary sector, that too of the rudimentary nature with small land holdings due to continuing fragmentation of land over time as a consequence of population pressure.

There has been a changing agricultural scenario with some vital inter-related issues like land-use, land relations, land reforms, landless labourers emerging. The politics of land reform constitute an integral part of the scenario. A rational solution of these issues will help agricultural development. Agriculture in both plain and in the hills of North East India has been undergoing some changes in the last two decades. This changing agricultural scenario is affecting population growth, migration and health of the rural people. We need to take an integrated approach to the multidimensional problems of changing agricultural scenario. The introduction in limited scale of modern technology in agriculture has brought about some changes in crop pattern and production in different areas of North East India where both settled and shifting cultivation are practiced. The social and economic background of the peasants and the ecological characteristics have to be given due weightage for understanding the degree of change and sustainability of gainful agriculture.

North East India Council for Social Science Research, the premier social science research organisation of North East India held a two-day seminar on Changing Agricultural Scenario in North East India on 12-13 December, 2004. His Excellency Mr. M.M. Jacob, Governor of Meghalaya, inaugurated the seminar. We thank him for his help and patronage.

Professor K. Alam, Professor T.B. Lahiri, Mr. S.K. Tewari, Professor B.C. Bhowmick, Professor A.C. Mahapatra and Professor N.N. Bhattacharya presided over the academic sessions. Professor Keya Sen Gupta, Dr. Sutapa Sen Gupta and Dr. Sushmita Das were rapporteurs of the seminar. We would like to thank them with our gratitude.

We take this opportunity to thank profusely Mr. C. Pynghope, Mr. S.K. Tewari, Mr. B.K. Dev Varma, Mr. P.C. Chakraborty, Mr. S.S. Gupta, Mr. K.L. Tariang, Mr. B.K. Panda, Mr. S.K. Sen and Mr. B. Dutta of the administration and Government of Meghalaya for their support and help to make the seminar a great success.

27th January, 2005

R.P. ATHPARIA,
Joint Secretary,
North East India Council,
For Social Science Research,
Shillong

CONTENTS

LIST OF TABLES

LIST OF APPENDIX TABLES

LIST OF FIGURES

LIST OF CONTRIBUTORS

Agrawal, K.N., Scientist SS (FMP), Division of Agricultural Engineering, ICAR Research Complex for NEH Region, Umiam, Meghalaya

Banerjee, Amalesh, Jadavpur University, Kolkata

Barua, Upala, Cotton College, Guwahati

Bhattacharjee, R.P., Joint Director (RD), Government of Arunachal Pradesh, Itanagar

Bhowmick, B.C., Professor, Department of Agricultural Economics, Faculty of Agriculture, Assam Agricultural University, Jorhat

Borah, B.C., Principal Scientist, NCAP, New Delhi

Borthakur, N., Associate Professor, Assam Agricultural University, Jorhat

Chakraborty, Gorky, Teachers Economics in Doom Dooma College, District Tinsukia, Assam

Chakravarty, D.N., Retd. Dean, Assam Agricultural University, Jorhat

Daimari, T. Chand, Head, Department of Geography, Union Christian College, Barapani, Shillong

Dutta, P.C., Department of Statistics, S.S. College, Hailakandi, Assam

Goswami, M., Lecturer, Department of Geography, Morigaon College, Morigaon, Assam

Goswami, Pranay Jyoti, Selection Grade Lecturer, G.C. College, Silchar, Cachar, Assam

Hazarika, C., Associate Professor, Department of Agricultural Economics, Assam Agricultural University, Jorhat

Jasmin Nilofar, Lecturer, Department of Anthropology, Morigaon College, Morigaon, Assam

Mazumder, Sukhendu, Scholar, C/o 99 APO

Mipun, B.S., Professor & Head, Department of Geography, NEHU, Shillong

Mishra, L.K., Division of Agroforestry, ICAR Research Complex for NEH Region, Umiam, Meghalaya

Mohapatra, A.C., Professor, Department of Geography, North-Eastern Hill University, Shillong.

Neog, A.K., Senior Member of the Indian Economic Service

Ngachan, S.V., ICAR Research Complex for NEH Region, Manipur Centre, Lamphelpat, Imphal

Pal, P.P., ICAR Research Complex for NEH Region, Umroi Road, Umiam, Meghalaya

Pradhan, B.C., Department of Statistics, S.S. College, Hailakandi, Assam

Rao, P.P. Nageswara, Deputy Director, North-Eastern Space Applications Centre, Shillong, Meghalaya

Saharia, Umesh Chandra, Selection Grade Lecturer, M.D.K.G. College, Dibrugarh

Sahu, B.P., Lecturer, UACE, NEHU, Shillong

Saikia, Shabeena Yasmin, O.K.D. Institute of Social Change and Development, Guwahati

Sarma, B.K., ICAR Research Complex for NEH Region, Umroi Road, Umiam, Meghalaya

Sarma, P.C., VAS, AH & Vety. Department, BCPP Scheme, Guwahati

Satapathy, K.K., Principal Scientist (SWCE), Division of Agricultural Engineering, ICAR Research Complex for NEH Region, Umiam, Meghalaya

Sen, Alok, Cachar College, Silchar

Sharma, P.J., Lecturer, Department of Geography, Morigaon College, Morigaon, Assam

Singh, R.K., Scientist (SWCE), Division of Agricultural Engineering, ICAR Research Complex for NEH Region, Umiam, Meghalaya

Singh, Rajmani, Manipur University, Imphal

Singh, S.B., ICAR Research Complex for NEH Region, Manipur Centre, Lamphelpat, Imphal

Singh, S.B., ICAR Research Complex for NEH Region, Umroi Road, Umiam, Meghalaya

INTRODUCTION

The state of agriculture in North-East India along with its potentialities and development thrust has been an inseparable part of the planning process over the years, a fact which initiated administrative measures for necessary impetus and some alternative strategies, wherever needed, for productivity and diversification. The North-Eastern region's identity is characterised by diverse geological, physiographic and agro-climatic features and except in some parts of Assam, Manipur and Tripura, the landscape is not alluring for cultivation. As a result, agro-climatic advantages which generate potentialities for cultivation of various types of agricultural and horticultural crops are overshadowed by drudgeries associated with field work. In addition, certain extreme agro-climatic conditions such as high rainfall, high humidity and low winter temperature also negate developmental efforts.

Today, agricultural economy of the region is on the brink of backwardness and stagnation aggravated by infrastructural bottlenecks, institutional weaknesses and technological lag. The factors which traditionally hampered agricultural productivity whether in the plains or in the hills have remained, by and large, unaddressed. No wonder, the persistence of traditional subsistence economy and rural poverty go hand in hand in most parts of the region, some of which are further affected by intra-regional disparities. It is difficult to ignore the fact that when governmental commitment to support agricultural process was at its height, the North-Eastern States could not rise to the occasion to give impetus to productivity and alternative choices. This was particularly true in the yester years of planning and nation building when the beginning could have been made in eventually releasing the economy from the vortex of stagnation. Now with the ascendancy of neo-liberal state, any expectation of massive state support has receded to the background. Nevertheless, agriculture remains the focal point of Indian planning. A growing awareness exists in different States of the region on the imperative need to overcome the impediments to agricultural development and make the economy self-reliant and vibrant so as to reduce its dependence on the rest of the country. However, a clear strategy is yet to evolve. As a first step, a macro-management approach through integration of 27 Centrally Sponsored Schemes has to be given a fair trial. In the changing scenario, agricultural development and food security are closely interlinked and in the ultimate analysis there is no better option than sustainable agricultural development.

Out of 2.55 lakh square km of the region, as low as 0.33 lakh square km is engaged in agriculture which forms 12 per cent of the total area. At the same time, of the 10.6 million main workers in the North Eastern economy, about 8 million are directly engaged in agricultural pursuits as cultivators and agricultural labourers. It is revealing that of the 22.5 million hectares of the region where land-use pattern is recommended, only 16.5 per cent (3.72 million hectares) is the net sown area for the purpose of cultivation. Significantly, a considerable large area of the region comes under shifting cultivation and the total number of families practicing shifting cultivation is about 4.4 lakhs. It may be stated that cropping intensity is 1.4 and the gross cropped area turns out to be 5.2 million hectares. In assessing the region's agricultural productivity, what is of great importance is the fact that the area covered by irrigation is negligible and it has been estimated that only 16 per cent of the gross cropped area is irrigated. The rest however remains embedded in the gamble of monsoons. At the same time, consumption of fertilisers is unusually low which is about 9.5 kg per hectare of gross area sown as against 67 kg per hectare at the national level.

In the North-Eastern Region, the share of agriculture in Gross State Domestic Product (at constant 1993-94 prices) varies from 20.17 per cent in Manipur to 30.5 per cent in Assam which undoubtedly corroborates the prevalence of subsistence economy. A fact which makes the situation so obvious is the pre-dominance of small and marginal farmers (about 80% of total holdings). Situation is worse in Assam where 83.12 per cent of total holdings are small and marginal operating upon 44.33 per cent of holdings. Inevitably, there is low agricultural productivity. An example from Assam shows that the yield of foodgrains per hectare in 2001-02 was 1,465 kg compared to 4,040 kg in Punjab.

According to one estimate, nearly one-sixth of the rural families in the region is landless and nearly one-half of the farmers including the landless hold less than ten bighas per family. It is well known that the average size of holdings is an important variable for agricultural sector. In the prevalent situation, mechanisation of agriculture through adoption of modern farming practices is ruled out. Further, roughly 28 per cent of the region's area is plains and the remaining 72 per cent is covered by mountain plateau and hills. At the same time, in the promotion of agricultural development land reforms initiated in the plains have so far been half-hearted, the hills remaining out of its purview out and out.

All this has had the cumulative effect of perpetuating low productivity giving no scope whatsoever for commercial surplus. Each State of the region has deficiencies peculiar to its terrain, fragility, inaccessibility and marginality of hill and mountain ecosystem coupled with traditional land-use system. In fact, six agro-climatic zones form the primary basis for development of agriculture in the region. Hence, a concerted move at the regional level may have the desired effect of ushering a common orientation to land-use and cropping pattern in the region. It is generally acknowledged that there are two ways to increase the production: increase in either productivity or cropping intensity. However, intensive farming with yield improvement strategies, crop diversification, development of cash crops and high valued crops hold the initial key to development of agriculture. Any such effort has to be supported obviously by institutional credit support, marketing facilities and input delivery. Further the abundance of ground water can easily facilitate a viable irrigation mechanism to guarantee higher yield. In fine, the emphasis has to be on ensuring a higher quality of life through maintaining a balance between productivity function and conservation activities.

This book is the outcome of joint efforts by academicians, planners and agricultural scientists to highlight the agricultural potentialities of the region against the backdrop of half-a-century of planning.

BIMAL J. DEB
B. DATTA RAY

1

WTO AND AGRICULTURAL DEVELOPMENT IN BACKWARD REGIONS

— A.K. NEOG

Introduction

Agricultural development policy in many countries of the world has been rooted in maintaining food security and achieving self-sufficiency. As a result, agriculture was virtually excluded from the General Agreement of Tariffs and Trade (GATT). It, however, came under the main disciplines of the World Trade Organisation (WTO) under Uruguay Round (1986-94) of the GATT. With the intention of aligning agricultural trade rules with those applying to trade in other goods, the Uruguay Round negotiators agreed that all import barriers, other than those in place for health and safety reasons, should take the form of transparent tariffs. Before agreeing on tariff reductions, all border measures had to be converted into their tariff equivalents, a process known as "tariffication." The Agreement on Agriculture (AoA) forms a part of the final Act of the Uruguay Round of Multilateral Trade Negotiations, which came into force on 1st January, 1995.

The basic objectives of the WTO Agreement on Agriculture (AoA) are to (i) establish a fair and market oriented trading system, and (ii) provide for substantial progressive reduction in the support and protection to agriculture resulting in correcting the distortions and preventing restrictions in world agricultural markets. In other words, the AoA aims at improving global trade, raising prices of agricultural products and ensuring higher living standard for farmers. The three main elements (also known as pillars) of the Agreement are market access, domestic support and export competition concerning the agriculture sector. A familiarity with these elements is necessary to comprehend the Agreement.

Market Access

Under the Agreement, market access has two sub-elements, viz., (a) tariffication of all non-tariff barriers, and (b) setting up of a minimum level for imports of agricultural products by member countries as a share of domestic consumption.

Tariffication implies replacement of non-tariff trade barriers [like quantitative restrictions (QRs), export and import licensing, etc.] by tariff measures in order to provide the same level of protection. Under the AoA, the QRs were to be withdrawn over a 6-year period ending 31st March, 2003. Tariffs, resulting from this tariffication process together with other tariffs on agricultural products, are to be reduced by a simple average of 36 per cent in the course of 6-year by the developed countries, and by 24 per cent during 10 years by the developing countries.

'Tariff binding'. is an important aspect of WTO regime. It is a commitment made by member countries not to raise particular tariffs above a specific or bound level (i.e., ceilings on tariff rates). As against this, applied (tariff) rates are rates of import duty which are actually applied. Applied tariff can be raised at any time provided they do not exceed bound rates.

Beginning in 1994, trade liberalisation was extended to agricultural products in India. There are now few restrictions on exports. Restrictions, however, remain on import of grains, oil seeds, and edible oils.

India has bound all its agricultural tariffs; its binding tariffs range from 0 per cent on primary agricultural commodities to 150 per cent for processed items, and edible oils upto 300 per cent. India can raise its applied rates in case of any surge in imports. However, according to estimates made by the Planning Commission, the weighted average import duty rates in Indian agriculture has gradually declined from the level of 47 per cent in 1991-92 to 14 per cent in 1997-78. Thereafter, it rose to 24.4 per cent in 1999-2000 and further to 57.7 per cent in 2001-02, due to imposition of various surcharges.

It may be mentioned that the tariffs are subject to progressive reduction commitments, except for rice and some staples that are subject to minimum access commitments, called Tariff Rate Quota (TRQ). TRQ is a measure under which a product is subject to a Most Favoured Nation (MFN) tariff, but a certain quantity (Quota) is admitted at a lower or zero tariff.

Domestic Subsidy

Commitments on domestic subsidy are expressed in terms of "Total Aggregate Measurement of Support" and "Annual and Final Bound Commitment Levels". The former refers to the sum of all domestic support or subsidy provided to agricultural producers (calculated on product-by-product basis), including the support for basic agricultural products (e.g., procurement price), non-product specific support (e.g., subsidies on fertilisers, power, credit, etc.) and all other supports to agricultural producers. The "Annual and Final Bound Commitment Levels" refer to the maximum support permitted to be provided during any year of the implementation period thereafter. WTO commitment requires a 20 per cent reduction in "Total Aggregate Measurement of Support" by developed countries over a period of 6-year, 13.3 per cent by developing countries over 10 years, and no reduction by the least developed countries.

The thrust of the domestic support commitments is on shifting away from trade-distorting measures and agricultural policy towards 'openness', transparency and competition. They are largely aimed at the developed countries where the levels of domestic agricultural support have accelerated in the recent years.

Under the Agreement, there are, however, provisions for some exemptions like government assistance on research, disease control, training, extension, advisory services, food security, domestic food aid, environmental programmes, regional development programmes, direct payment to producers like participation in income insurance and safety nets, relief from natural disasters (known as 'green box' subsidy in WTO jargon). Green box subsidies are permitted; they have no trade-distorting effect, hence non-actionable under WTO. They are for production restructuring and direct payments to producers. Similarly, there are 'amber box' subsidies (subsidies that distort production and trade) which are to be reduced. They include measures to support prices, or subsidies directly related to production quantities. These are subject to limits: "de minimus" minimal supports are allowed, i.e., total support of 5 per cent of agricultural production for developed countries and 10 per cent for developing countries. 'Blue box' subsidies are the "amber box with conditions"—conditions designed to reduce distortion. Any amber box support that requires farmers to limit production come under blue box. There are also exemptions for developing countries given under the Special and Differential (S&D) treatment (i.e., the S&D box), including those in Article 6.2 of the AoA. Examples are investment subsidies and agricultural input services generally available to income and resource poor producers.

Export Subsidy

Export subsidies are prohibited under the WTO. Members are required to reduce the value of direct export subsidies to a level of 36 per cent below that during 1986-90 base period in the course of a 6-year implementation period. The quantity of subsidised exports is to be reduced by 21 per cent over the same period. For developing countries, the reductions are two-thirds of those of the developed countries over a 10-year period, and there are no reductions for least developed countries. Export subsidies on agriculture are virtually non-existent in India and, hence, reduction is not relevant for the country.

The developing countries have a special ten-year period for protection from action by developed countries under the 'Peace clause' given in Article 13 of the AoA. It restricts the right to take action that might otherwise be possible against domestic support measures and export subsidies in agriculture. Under this clause, no countervailing (duty) action can be taken against permitted measures. However, this clause is scheduled to expire by 31st December, 2003. With its expiry, the developing countries can impose countervailing duties on import of developed country products that are heavily subsidised causing injury to domestic producers. In the Draft Ministerial Text at Cancun, there was a proposal for extending the validity of the peace clause.

Implication of WTO agreement on agriculture for India needs to be viewed basically in terms of (a) opening up of markets and facilitation of exports of our agricultural products, and (b) degrees of freedom to pursue domestic policies.

India's Position

Domestic support to agriculture in India continues to be provided through subsidised inputs and support prices for rice, wheat, oil seeds, etc. It is worth mentioning that in the 1986-88 base period for the determination of commitments under the AoA, India's Aggregate Measure of Support (AMS) for each product was below the 10 per cent 'de minimus' level limit. Hence, India has no total AMS reduction commitment under the Agreement. Nevertheless, India is required to make regular annual notification to the WTO on its domestic AMS, and on its direct export subsidies.

India submitted its negotiating proposals in respect of agriculture on 15th January, 2001, covering domestic support, export subsidies, market access, and S&D treatment including food and livelihood security. The proposal calls upon the developed countries to reduce tariff peaks and eliminate tariff escalation, simplification of administration of tariff rate quotas, prohibition of dumping, elimination of export subsidies and substantial reduction of their domestic support. It also calls for abolition of 'peace clause' for developed countries.

Doha Declaration

It is to be noted that in respect of agriculture, Doha Ministerial Conference in November 2001 agreed to substantially reduce domestic support, reduce (with a view to phasing out) export subsidies, substantially improve market access, and stated that S&D treatment shall be an integral part of all elements of the negotiations, and non-trade concerns, such as animal welfare, shall be taken into account. Further negotiations are to be concluded by 1st January, 2005.

Pre-Cancun Exercises

Negotiations on agricultural tariff reduction was to follow in the light of formula approach. The widely discussed formula were the Swiss formula, Uruguay round formula and Harbinson proposal. The Swiss formula was proposed in the Seventh Round (1973-79) of GATT held at Tokyo. It envisages steeper cuts on higher tariffs. The Uruguay Round (1986-94) formula envisages linear reduction in tariff lines which call for equal percentage for all, irrespective of the prevailing tariff rates.

The Harbinson proposal, named after Stuart Harbinson, the Chairman of the WTO negotiating group on agriculture takes a middle line. For developed countries, it proposes average tariffs cuts of 60 per cent on bound tariffs with above 90 per cent binding level; a 50 per cent cut on bound tariffs between 15 and 90 per cent; and a 40 per cent cut on bound tariffs below 15 per cent. For developing countries and for non-strategic products, it proposes average cuts from bound rates of 40 per cent for tariffs above 120 per cent; a cut of 35 per cent for tariffs between 60 per cent and 120 per cent; a cut of 30 per cent for those between 20 per cent and 60 per cent; and an average reduction of 25 per cent in tariffs below 20 per cent. Under the proposal, reductions are to be implemented in equal instalments over five years by the developed countries and over ten years by developing countries. Harbinson proposal implies substantial tariff cuts by US and EU.

Cancun Ministerial

The Fifth Ministerial Conference (coming once in every two years) of WTO was held at Cancun in Mexico from 10th to 14th September, 2003, participated by 146 members, where liberalisation of international trade in agriculture was an agenda item, *inter alia*. Arriving at Agreement on the item envisaged reduction of domestic support and phasing out of export as per Doha declaration.

It may be mentioned that the Cairns Group (Australia, New Zealand, etc.) has been pushing for steep cut in subsidy as well as tariffs since they are strong in agriculture. But the rich countries like US, EU are reluctant in subsidy reduction. On the other hand, average agricultural tariffs in rich countries are several times higher than those on their manufactured goods. Tariff peak is as high as 1000 per cent, for example, on rice in Japan.

India, in alliance with other developing countries like Argentina, Brazil, China, South Africa, etc. (a group of about 22 countries), presented a plan in the Ministerial envisaging more subsidy cuts and tariff reductions by the rich countries compared to poor countries, thereby removing trade barriers and making a fair deal for farmers.

The core issue in Cancun was cut in farm subsidies which was in pursuance of commitment at Doha. But the US and EU were rigid on their stand against subsidy reduction. Agreement was "too near, yet so far" at Cancun. Hence no agreement was arrived at. Outstanding issues, *ipso facto*, get postponed. This implies that the rich countries will continue to maintain large domestic support to farmers and for export.

Views on Cancun Outcome

Several views have been aired about the impact of Cancun failure on India. Professor T.N. Srinivasan of Yale University in an article published in *Financial Express* dated 4th September, 2003 has expressed—"we should aspire to be viewed by the rest of the world as an economy from which it can buy a range of quality products at competitive prices, and to which it can sell its products in a stable market and policy environment with few barriers". WTO is a forum for continuous multi-lateral negotiations. It does not prevent a member from autonomous/unilateral liberalisation. Prof. Srinivasan also says, "since unilaterally opening our markets to international competition is in our own interest regardless of whether our access to markets of others is enhanced, getting such enhanced access in return in negotiations is a bouns". Where bilateral deals are feasible, India should start negotiating the same. However, net gains from negotiations would be minimal *vis-á-vis*, our competitors if our infrastructure like irrigation, are not strengthened, law and order are not improved, and other domestic constraints on international competitiveness are not addressed. This needs agricultural reform.

Another view expressed is that Cancun is merely a mid-term review of the on-going negotiations under Doha agenda. There would be no harm on Indian economy, particularly agriculture, as our primary objective is to safeguard our farmers' interest. Indian agriculture is largely competitive; net agricultural imports have been almost constant and has nothing to gain in agro-exports. However, this view is a *status quo* onc and suffers from myopia. They consider trade as a 'zero-sum' game. It may be mentioned that a globalised free market is premised on the principle of dynamic comparative advantage. In a borderless world, each individual economy is supposed to gain from global trade by focusing/specialising on areas where it has inherent advantages. Hence, trade is a 'positive sum' game in which everyone can win. WTO must ensure that global trade is equitable.

An Evaluation of Trade

It is necessary to take a look at India's overall foreign trade to assess its strength, weakness, opportunity and threat (SWOT). According to WTO's World Trade Report (2003), in world merchandise trade in 2002 India ranked 30th in the list of exporting countries with a share of 0.8 per cent and 24th in the list importing countries with similar share of 0.8 per cent. This shows an improvement from India's 0.5 per cent share in World export in 1990 and 0.6 per cent in 1995.

Role of WTO is envisaged to enhance the degree of openness of an economy. One measure of openness is the ratio of a country's exports to the sum of its GDP at current prices and its imports. In the case of India, this ratio had been about 8.16 per cent in 1993-94, 8.59 per cent in 1996-97, 9.47 per cent in 2000-01, 8.94 per cent in 2001-02 and 9.96 per cent in 2002-03. This shows that during the liberalisation era India's openness has been rising.

Looking at the share of agriculture and allied products in India's total exports, it is observed that it has steadily come down from about 20.5 per cent in 1996-97 to 12.26 per cent in 2002-03. It may be mentioned that agricultural commodities suffer from low income elasticity and high price elasticity of demand. However, their share in total imports have risen marginally from 3.5 per cent in 1996-97 to 4.31 per cent in 2002-03. Their ratio of exports to imports have fallen from 5.01 in 1996-97 to 2.43 in 2002-03, though value of exports of agricultural and allied products are higher than their imports.

Global Agricultural Scenario

Global agriculture is characterised by heterogeneity. Agricultural policy scenario in global context is best presented by World Bank in its report "Global Economic Prospects 2004: Realising the Development promise of the Doha Agenda". Some of its stylised findings are furnished below:

(i) Trade in agriculture is important to the world's poor but agricultural policies have often worked against the poor

Developed country agricultural reforms have been modest despite the inclusion of agriculture under the WTO. The result of their policies has been overproduction and price declines in many commodities, reducing opportunities for many developing countries to expand exports and penalising the world's poor. During the 1990s, the growth of developing country agricultural exports to industrial countries slowed as exports to other developing countries accelerated.

The benefits of global liberalisation in agriculture through elimination of all border barriers and subsidies are estimated to be very large for industrial and developing countries alike, topping $ 350 billion for the world. With liberalisation, agricultural production would marginally shift from North to South, and the highly depressed world prices for many commodities would increase by 10 to 20 per cent each for cotton and ground nuts, 20 to 30 per cent for dairy products, 33 to 90 per cent for rice, and 20 to 40 per cent for sugar.

(ii) Agriculture is the livelihood of the world's poor

Using the $ 1-a-day measure of poverty, the Report says that most of the world's poor live in India, China and other lower-middle-income countries.

(iii) Poverty is more common in rural areas

Poverty in rural areas is closely correlated with distance to local and national markets. Without improvements in the functioning of local and national markets, economic gains for the poor may reach only one-fourth of their potential. Hence, improvements in trade policies alone may not be sufficient to restore sustained growth in the agricultural sector without better transport infrastructure and other reforms.

(iv) Farmers in industrial countries earn above average incomes

By implication, developing country farmers will not be in a position to compete with them. It, therefore, calls for level playing field.

(v) High border protection in rich countries frustrates development

Border protection in rich countries continues to be high, non-transparent, and anti-development. Tariff peaks as high as 500 per cent confront imports from developing countries. Tariffs also increase by degree of processing; creating a highly escalating tariff structure that limits access for processed foods.

(vi) Agricultural subsidies in rich countries distort global income distribution

The subsidy programmes prominent in current food and agriculture policy are not targeted to keeping small and marginal farms in business but instead provide hefty rents to large farmers. Nor are current production-based policies effective in achieving their various other objectives (such as environmental sustainability and rural development). By increasing land prices they also lead to the creation of larger farms and the elimination of small farms. Their spillover effects on global markets, and on other countries, are large and negative. When subsidies depress prices the impacts in poor countries can be severe.

(vii) Global markets are important for sustained income growth in developing countries

For countries with a relatively small urban population, agricultural exports can produce faster growth than can domestic market demand. In such cases, the international market provides growth opportunities. Export growth contributes significantly to the growth of non-export agriculture by providing cash income that can be used to modernise farming practices. For those leaving the farm, growth and modernisation of agriculture create jobs in agro-processing and marketing.

(viii) Product trends in exports differ

Static markets in industrial countries for traditional developing country products such as coffee and tea seem to have contributed to declining import growth rates in developed countries, as did the decline in their GDP growth rates, combined with low elasticity of demand.

The protection rates for food processing in industrial countries are extremely high, far above those of any other manufacturing sub-sector. The developing countries have not gained market share in food processing.

(ix) The evolving structure of trade is towards non-traditional products with lower rates of protection

Export product groups that showed significant declines during, 1980-81 and 2000-01, were grains (14.3 to 9.5%); coffee, cocoa and tea; sugar and sugar products and textile fibres—all of which are among the traditional exports of developing countries. The declines were caused by a combination of price declines, low demand elasticities; and in the case of sugar, grains, meats, and milk by high rates of protection and expanded production in industrial countries. Export of fruits, vegetables, cut flowers, fish, seafood, alcoholic and non-alcoholic drinks gained significantly during the period.

(x) Global agricultural protection is a bias against development

Protection in agriculture has taken different forms like tariff protection, subsidies, tariff peaks, Tariff Rate Quota (TRQs), tariff escalation, and opaque tariffs. These are often inter-linked. The entire inter-linked system of protection, even when used by other developing countries, is heavily biased against developing countries and against the world's poor.

It is envisaged that elimination of tariffs and subsidies on agriculture would lead to rise in product prices due to free play of market forces, and benefit the farmers. If the current level of tariff on a product is lower than the anticipated global price increase, then there may not be direct consumer benefit as the post-reform consumer price would be higher. However, higher market price would attract the farmers to produce more.

National Agricultural Scenario

The following are some of the main features of India's agricultural scenario and its performance:

(i) There are 1,155.8 lakh number of holding operating on 1,633.57 lakh hectares in India as per 1995-96 Agricultural

Census. Of these, 928.22 lakh number (80.31%) are small and marginal holdings operating on 588.43 lakh hectares (36.02%) area only.

(ii) As per provisional data of population census 2001, the number of cultivators in India stands at 1,276.00 lakh and agricultural labourers at 1,075.00 lakh.

(iii) As per Central Statistical Organisation (CSO), India's agriculture and allied sector accounts for 23.8 per cent of GDP at factor cost in real terms in 2000-01 (23.9% QE in 2001-02) which have gradually declined from 31.0 per cent in 1993-94.

(iv) While the overall agricultural growth during the last four decades has been more or less constant, the relative contribution of area expansion and of improvement in output per hectare has progressively shifted in favour of the latter. In the fifties, the major part of increase in output was from area expansion, at present almost the entire increase in output is from more intensive cropping of available land and from higher yields per hectare.

(v) There are vast inter-crop and inter-regional differences in performance, the fastest growth has been in the wheat growing regions of North-West India which have benefited from the advent of HYVs and well-developed irrigation facilities. Despite HYVs, growth has, however, not been impressive in the predominantly paddy growing areas, especially in East and North-East India.

(vi) While the inter-regional disparities in output per hectare have not widened, the disparities in terms of output per head of agricultural population has been widening progressively.

(vii) In general, productivity per hectare is increasing mainly on irrigated land; productivity of unirrigated land seems to be more or less stagnant in most of the States.

(viii) While HYVs and fertiliser use have spread, the level of yields achieved is, in general, considerably below the potential demonstrated to be attainable with proven technology.

(ix) Indications are that the incremental returns on the irrigation-fertiliser-improved seed package are declining.

(x) Though inequality in land distribution has not worsened, there is sharp decline in average size of holdings on account of rise in population. The average size of holdings in 1990-91 was 1.57 hectares which declined to 1.41 hectares in 1995-96. Disguised unemployment persists in Indian agriculture.

India occupies second position (China first) in world agriculture during 2000 in the production of rice (22.4% of world) and wheat (12.8%). However, India's yield per hectare of rice is lower at 3,008 kg. compared to Egypt's 9,086 kg. (highest in the world) or China's 6,234 kg. Likewise, India's wheat yield is 2,777 kg. per hectare compared to UK's 8,006 kg. (highest) or China's 3,729 kg. In pulses, India occupies first position (23.6%). In groundnut production India's position is second (17.1%), next to China. However, its yield per hectare is lower at 859 kg. compared to China's 3,329 kg. (highest). The huge gap in yield reflects not only agro-climatic factors but also the differences in technology.

Regional Economic Scenario

India ranks first in the world in the production of tea, producing about 25 per cent of world output in 2000. Of particular interest to the regional economy of the North-East (NE) India is the export of tea. Share of tea in India's total exports was about 0.87 per cent in 1996-97, 1.62 per cent in 1998-99, and it stood at 0.64 per cent in 2002-03. Total export of tea in India has increased annually by 2.3 per cent at compound growth rate during 1996-97 and 2002-03. Its import started from 1999-2000 (US$ 5.9 million). However, imported tea was US$ 25.7 million in 2002-03 compared to its export at US$ 336.0. With the removal of QRs from 1-4-2001, tea can be imported from any country without any licence. Tea import is accelerating.

WTO is universal. No region of a member country of WTO can remain out of its orbit. As per 1995-96 agricultural census, the total number of holdings in the seven States (Arunachal Pradesh, Assam, Manipur, Meghalaya, Mizoram, Nagaland and Tripura) has been 36.06 lakh (i.e., 3.12% of all-India) with operating area of 48.55 lakh hectares (i.e., 2.97% of all-India).

North-Eastern region has the following unique features:

(i) Share of agriculture in Gross State Domestic Product (at constant 1993-94 prices) was 30.5 per cent (highest in the region) in Assam in 2001-02, in Manipur 20.1 per cent (lowest in the region) in 2000-01. The growth of gross agricultural SDP was 0.2 per cent in 2001-02 for Assam; 2.8 per cent and 2.9 per cent respectively for Manipur and Tripura in 2000-01, while 8.8 per cent in Arunachal Pradesh but negative for Meghalaya. For the other States data are not available.

(ii) Agricultural dualism with highly modern tea plantation co-exists with subsistence farming in the peasant sector.

(iii) Pre-dominance of small and marginal farmers (about 80% of total holdings) operating on about 38 per cent of the area compared to all-India average of 36 per cent area. It can be mentioned that in Assam as per 1995-96 agricultural census, 83.12 per cent of total holdings are small and marginal, operating upon 44.33 per cent of total operational holdings.

(iv) Abundance of surface and ground water potential. However, a significant part of the region is flood-prone (compared to all-India figure of 12 per cent land vulnerable to flood), and all the States including Assam suffer from major soil erosion/ land slide problems.

(v) The region is characterised by low agricultural productivity and input use. In Assam, the yield of foodgrains per hectare in 2001-02 was 1,465 kg. compared to 4,040 kg. in Punjab. Areas covered under irrigation in 1999-2000 was mere 7.6 per cent compared to 96.6 per cent in Punjab.

(vi) The entire region is characterised by weak agricultural institutions like credit, marketing, input delivery.

Within the region, Assam accounts for 74.40 per cent of the holdings with operational area accounting for 64.63 per cent of area. Most of the farms are marginal and small. In the hill areas, *jhum* (shifting) cultivation is widely practised. While tea cultivation follows the market-oriented agricultural model, the model followed in the case of crop farming, by and large, is the traditional subsistence. Around 50 per cent of the area under paddy during 1995-96, which is the main crop in Assam, has been said to be covered under High Yielding and improved varieties.

As per 2001 census, the number of cultivators in the region stood at 60.83 lakh (i.e., 4.77% of all-India) and agricultural labourers at 19.42 lakh (i.e., 1.81% all-India). The number of cultivators in Assam stood above 37.42 lakh (i.e., 61.52% of the region) and agricultural labourers 12.90 lakh (i.e., 66.43% of the region). These two categories together account for 52.65 per cent of total workers in Assam. Hence, within the region, WTO has got more implication for the livelihood of workers in Assam.

According to Planning Commission's data, average value of land yield (value of output divided by the cropped area) which is also a measure of agricultural productivity has moved as under for the selected States (given for reference) during the given period. It may be noted that value tends to vary depending on the type of crops grown and their prices. Hence average yield will be higher in respect of high value crops like plantation, horticulture and commercial crops than common foodgrains.

Table 1.1: Average Value of Land Yield

State / Period	Average Yield (Rs./Hectare) at 1990-93 Constant Prices	
	1980-83	1992-95
Kerala	12,333.85	15,625.96
Punjab	9,707.65	13,597.22
West Bengal	5,943.81	9,958.45
Assam	6,906.69	8,196.82
All-India	5,090.42	7,388.05

Source: Planning Commission, *Tenth Five Year Plan* (2002-07), Vol. III, p. 78.

As per the data, yield per hectare is the highest in Kerala (followed by Tamil Nadu); Punjab and Haryana occupies third and fourth position respectively. Though Assam's figures are higher than all-India, its increase (1.44% compound growth rate) in productivity during 1980-95 has been much lower than all-India growth rate of 3.15 per cent. In fact, Assam's growth rate is the second lowest. West Bengal registered the highest growth rate at 4.39 per cent followed by Haryana, 4.13 per cent. The Plan document has also stated that the States in which agricultural production has increased significantly are the States in which there has been a rapid decline in poverty. In view of this, it can be said that success or failure in agriculture sector as a result of the WTO regime has got serious economic implications.

The above figures need some comments. The annual yield per hectare of Rs. 8,196.82 (at constant price) translates to Rs. 683.07 per month for Assam. When divided by an average family size of five, the monthly earnings per head stands at Rs. 136.61. For a two-hectare farm in Assam, which is preponderant, the monthly earnings per head is Rs. 273.23. This shows the subsistence character of agriculture in the State. Low farm productivity results in high unit costs of food, and lower farm income.

Suggestions

Agricultural development is basically a function of six elements (6 'I's), viz., Institutions (including land tenure, R&D, training, marketing, credit, stock exchanges in agricultural goods, etc.), Infrastructure (including power, irrigation, cold storage, transport, etc.), Investment, Inputs (seed, fertilisers, etc.), Incentives and Information (on technology, prices, costs, quality, etc.). Agricultural reform programme should take into account these elements. Agriculture is subject to diminishing marginal productivity of labour and returns to scale. Programmes to invest in 'public goods' such as Research and Development (R&D), human resources, improving safety nets, which are also permissible under the WTO, enhance productivity, reduce farm risk and uncertainty. Crop diversification, which is basically a function of research, helps in minimising farmer's risk and uncertainty.

Agriculture is a State subject under the Constitution of India. Expansion of trade opportunities is an important ingredient of trade liberalisation. Policies assuring the small and marginal farmers their access to the above elements can facilitate their participation in markets and in exploiting export opportunities.

Poor infrastructure is the Achilles heel of the North Eastern region. The Green Revolution in Punjab is basically due to its sound infrastructure, particularly in irrigation. Its highest yield of foodgrains per hectare is attributed to irrigation, which covers 96.6 per cent of the cropped area. Irrigation is, therefore, a *sine qua non.* It should come from the public sector and/or public-private partnership.

In a peripheral economy, individual farmers under globally open economic system are highly exposed to global shocks and swings. The small and marginal farmers should form Associations so that their voices about their stakes are better heard in political circles under the WTO regime and can strike better bargains for protection.

In the hill areas where there in no cadastral survey, the land is community-owned and institutional credit is absent, 'contract farming' can offer a solution in commercial production, supply chain management and marketing. Contract farming is an agreement between farmers and sponsors, i.e., government bodies/corporate/processors/entrepreneurs to coordinate and promote production and marketing. The traditional farming is based on the strategy of "supply creates its own demand", *a la Say's law.* Contract farming reverses this as production is demand/market driven. Farmers have to produce as per sponsor's requirement; the latter also supplies inputs and credit, and procure the products at fixed/contract price. Punjab Agri Export Corporation Ltd. is a pioneer in contract farming. In this respect, partnership between farmers' association and MNCs, share holding companies in land can be established. However, these require institutional reforms, particularly in customary and land laws. Wherever feasible, farmers in order to optimise returns from tiny plots, should switch over from subsistence farming to high-valued crops like vegetables, fruits, medical and aromatic plants, floriculture. These products are labour intensive and require less land per unit of output. They are suitable for land-scarce, labour-abundant economy and hill areas. These are high-valued, income elastic products and have global markets. However, they may need support of modern technology, intensive care, cold storage and quick marketing.

The NE region is surrounded by foreign countries. This offers opportunity for border trade through which 'surplus products' can be exported. Export promotion industrial parks, border trade centres, etc. have already been identified or are in the process of establishment in selected parts of the region. Under the Exim Policy, forty-eight Agri-Export Zone were approved till July 2003 by the government. Such projects (one each), for ginger in Assam and for Pineapple in Tripura are coming up. Success of such projects lies on the 'partnership' envisaged among Governments, farmers, entrepreneurs and exporters. The other States should also come forward with proposals for Agri Export Zones to gain from competitive advantage so that exports from the region bloom.

The WTO is omnipresent. Opening of agricultural sector brings global competition, which is looming large everywhere, and which small and marginal farmers will also have to face. To transform the laggard agricultural economy into a vibrant one, the governments will have to provide infrastructure and usher in institutional reforms. The farmers have to raise their competitive strength through diversification and higher their productivity and yield per plot in order to seize the emerging opportunities.

REFERENCES

Baishya, P. *et al.* (ed.), *Development Issues of North-East India*, Lawyer's Book Stall, 1997.

Chattopadhyay, M. *et al.* (ed.), *Planning and Economic Policy in India*, Sage Publications India (P) Ltd., 1996.

Goswami, P.C. (ed.), *Agriculture in Assam*, Assam Institute of Development Studies, 1989.

Goswami, P.C., *The Economic Development of Assam*, Kalyani Publishers, 2003 (Revised by A.K. Neog).

Government of India, Department of Agriculture and Cooperation, *Agricultural Statistics at a Glance*, 2003.

Hoekman, B. *et al.* (ed.), *Development, Trade and the WTO*; World Bank, 2002.

Neog, A.K., "Intra-regional Trade and its Potentiality for North-East India" in Gauhati University, *Journal of Commerce*, Vol. V & VI, 1998.

Neog, A.K., "New World Trade Agreement" in D. Bhorali (ed.) *GATT Agreement or Dunkel Draft Treaty*, Mittal Publications, 1994.

Passah, P.M., *et al.*, *Agricultural Transition in the Hill Areas of the North Eastern Region* (4th Conference of the NEEA), published by B. Syiem, Shillong College, 2002.

Srinivasan, T.N., "India must stop being purely defensive in WTO," in the *Financial Express*, 4th September, 2003.

World Bank, Global Economic Prospects 2004: Realising the Development Promise of the Doha Agenda, 2003.

2

HAS AGRICULTURE CHANGED IN THE LAND OF SEVEN SISTERS?

— A.C. MOHAPATRA

Background

Decidedly changes do occur through time, for better or for worse. It is societal evaluation and judgement that morality can be assigned to events, whether such changes are good for people or not, whether the processes and their correlates desirable, or for that matter whether the process is progressive, moving the society to a higher plane of achievement. And how can we make such a judgement? In fact, the bases of such moral judgements are to be sorted out before we go for evaluation. The intention in this chapter is to address such bases of evaluation, and then to evaluate whatever is happening in the agricultural situation in the region.

First is the question of quantitative growth in agriculture. Decidedly this does not happen in agriculture in isolation from other sectors of the economy, nor without competitive influences from other regions of the country. But, nevertheless, higher growth, especially led by higher productivity of land (yield) is generally accepted as a better situation. This is principally, for two reasons: (a) Since, a very large proportion of the population are farmers, larger production and productivity is assumed to make the small farmers have access to food from their own labour and the little bigger farmers will have a better disposable income at their hands, therefore, increasing their economic and social welfare; (b) the region being chronically short of food in relation to what is locally produced, it provides a comfort level of regional food security.

Second relates to the question of distribution of benefits from growth. Two types of changes may occur: (a) changes in land (asset) distribution; e.g., higher incomes (whether from agriculture or other sectors) may change land distribution, through the 'hidden hand'; (b) the nature of product distribution, i.e., whether the increased availability of say food, reaches or meets the nutritional needs of the vulnerable groups of people. To think that food self-sufficiency of a State or a district solves the problem of the hungry and under-nourished is banal. Here, as Professor Sen argued, "access" to food is more important than "availability", unless we assumed a closed economy with no external trade from that district or State. Else, we have to look at a progressive policy regime by the State to make food accessible at lower prices to the most vulnerable classes, as probably is pursued now in India, with doubtful effectiveness.

Third relates to the question of rural employment (wage employment) emanating from increased disposable income at the hands of (at least) the larger farm holdings, the large farmers being willing to "deepen capital" with employment opportunity for the wage earners. This is not guaranteed. Indeed, if one looks at the rural income elasticity for consumer products, it is far higher than the response to employment. The rural rich would tend to splurge on either durable assets (including land transactions) or on current consumption rather than further invest in farming (Bharadwaj, 1975), which is especially true of poorer agricultural States.

Fourth is the question of rural poverty itself. The expectation that growing agriculture could bring down the level of rural poverty is never automatic without appropriate enabling public policies, like food subsidies to the poor, crop insurance and other social sector inputs. Indeed, experience of some developed agricultural States like Punjab, Andhra Pradesh and Karnataka, etc. indicate deteriorating condition of small and marginal farmers, land alienation, and suicides (due to increased debt burden and crop failures). On the other hand, subsidies on fertiliser, water and electricity have a built-in favourable bias to the rich farmers. These issues are well known. We are in any case, not very clear about higher agricultural growth and reduction in rural poverty, with possible exception of the States of Punjab and Haryana, where urban poverty exceeds rural poverty. Here, however, the data does not indicate who these urban poors are? Invariably these are the physical transfer of poor (destitution) from rural to urban areas, or migrant wage-workers from other States and areas. During the nineties, West Bengal has emerged with the highest agricultural growth, but has it led to reduction of rural poverty to an acceptable level?

In fact, there have been some changes in agriculture in the NE region, both in the plains as well as in the hill areas in the last two decades, though the pace of growth is far from satisfactory. Many strategies have been introduced and then discarded without very definite results and the experimentation has continued. Undoubtedly, the potential is great and it would be unthinkable for the region to progress without a resilient agriculture sector. But, it is more important too, to judge whether such growth has led to more equitable and desirable welfare for the bulk of the population aspiring for a better standard of living. Instead of a mechanical narration of the success or failure of the farming sector, the attention has to be focussed on the consequences of such growth or progress or whatever nomenclature one chooses to use.

Agricultural (Quantitative) Growth

It is generally perceived that growth is good, for the country, for the region and for the people. Agricultural growth is even better since a large part of the population is, invariably dependent on agriculture for their livelihood, especially in an agrarian society like ours. However, growth could be of total output resulting from either change in expansion/contraction of cultivated area or change in productivity (of land as well as labour). Change in cultivated area may arise out of: (a) actual area increases or brought under crop practices, (b) multi-cropping, i.e., increases in the number of times the same piece of land is cropped, and (c) interchanges of areas put to different crop practices due to various responses to prices and other non-monetary considerations. Land productivity changes are due to yield changes in different crops, which in economic terms is a rent growth, whereas the labour productivity changes is connected with the concept of value addition and can be linked with wage rates and change thereof.

On the other hand, changes (or growth) in agriculture cannot be viewed in isolation from changes in other sectors, like fertiliser or diesel oil (for energising private irrigation) and other such inputs whose availability and prices are important for farmers' decision to use them affecting output and yield in agriculture. Farmers' decision to sow some crops or not is not entirely dependent on his family's requirements but market conditions and even access to markets, either physical access or information. Similarly, public policies could effect farming practices like providing support prices to selected and strategic agricultural goods or other forms of subsidies, fiscal incentives, etc.

Finally, the farmer's own situation, his socio-economic condition, size of operational holding, access to water management (public irrigation), access to credit which undoubtedly have important influence on production and productivity in farming the practices. Even environmental conditions like flooding or droughts, precipitation and other factors would influence the quality of farming practice. Harvest of the crop and output may reflect a result of the complexities of technical, economic and socio-economic situation of the farmer placed in particular locales, environmental niches and the manner he may try to get the best out of the situation. Whereas his welfare is directly dependent on his harvest, he may have very little influence on the whole complex of realities, which affect the crop output.

When one looks at the agricultural changes in the North-Eastern region, one has to be conscious of these complexities within which the farmers operate, their specificities and peculiarities, the political economy in assessing what meaning these changes may imply for the welfare of the farmers. However, whereas at one level one wishes to understand the changes in the region as a whole, the complexities of environmental and societal conditions are diverse and thus, provide different opportunities as well as impose variegated constraints. There are the great river valleys of Brahmaputra, Barak and Imphal Rivers that largely produce the staple food, rice, in family held subsistence farms. On the other hand, the surrounding hills are home to myriad tribesmen, small and large, but nucleated into specific regions and niches who have a wide variety of farming practices ranging from rice-paddies in small valleys, to vegetable farms on slopes, sylviculture, horticulture, animal rearing and to shifting cultivation in many areas, which has great implication for the subsistence of the families concerned but has little implication in a rapidly expanding market economy. Indeed, little definite information is available at the State or regional scale of the areas cultivated under such practices or the total volume of products of commercial significance, except products like potatoes, ginger, and horticultural products like pineapples, oranges and now, apples which have appeared in the market, principally consumed within the region itself. So, essentially, by way of hard facts available are principally on rice and some other minor cereals and oilseeds. Ventures into animal farming, dairying, poultry and fishery are in very small scale and are of limited economic significance at the moment, since large volume of these products are in any case regionally imported in substantial quantities (over one-fourth of the quantity demanded) to meet the overall demands in the region.

Table 2.1 shows the State-wise figures for area, production and yield of rice between 1985-86 and 1997-98, a period of 12 years. Rice being the staple crop and food of the people, it is itself an indicator of the ground reality. The total rice output of the region improved from 3.986 million tonnes in 1985-86 to 4.848 million tonnes in 1997-98 at an arithmetic rate of 1.8 per cent a year, way below the growth of population (2.5+ per cent), which means the per capita availability of the regional produce has been on the decline during the reference period. Coupled with this, the overall demand on foodgrains in the region (around 175 kg. per head/year) of an estimated over 6.5 million tonnes, the deficit continues to increase with increasing dependence of the region for imports from other regions of India to the tune of nearly Rs. 2,000 crores either through the FCI mechanism or the open market transactions. Except the tiny States of Meghalaya (3.48%), Mizoram (3.25%) and Nagaland (1.67%), the contribution of area growth in rice has been negligible a fact that indicates exhaustion of easier potentials for expansion of rice farming and output rise accruing from expansion of area under rice.

However, area expansion arising out of intensive practices have not been exhausted, though irrigated rice double cropping potential is limited in the hill areas as scope for public irrigation works is limited. There exists, however, large such scope in the fertile valleys of Brahmaputra, South Assam and Imphal Valleys, which have been imbued to traditional single rice crops. In very recent times, in certain areas of Brahmaputra Valley energised private irrigation and

Table 2.1: Area, Production and Yield of Rice

(Area in '000 ha, production in '000 tonnes and yield in kg.)

States		Area, Production & Yield				Growth Rates per Year			
		1985-86	1990-91	1995-96	1997-98	85-86 to 90-91	90-91 to 95-96	95-96 to 97-98	85-86 to 97-98
Arunachal Pradesh	A	108.0	121.8	118.7	120.0	2.56	–0.51	0.55	0.93
	P	117.6	142.5	124.5	129.5	4.23	–2.53	2.01	0.84
	Y	1,088.0	1,170.0	1,049.0	1,079.0	1.51	–2.07	1.43	–0.07
Assam	A	2,464.1	2,490.0	2,503.4	2,489.8	0.21	0.11	–0.27	0.09
	P	2,846.6	3,270.2	3,390.0	3,382.9	2.98	0.73	–0.10	1.57
	Y	1,155.0	1,313.0	1,354.0	1,359.0	2.74	0.62	0.18	1.47
Manipur	A	164.6	157.4	154.4	157.9	–0.87	–0.38	1.13	–0.34
	P	332.5	274.2	331.8	351.7	–3.51	4.20	3.00	0.48
	Y	2,020	1,742.0	2,149.0	2,227.0	–2.75	4.67	1.81	0.85
Meghalaya	A	111.4	103.6	104.0	157.9	–1.40	0.08	25.91	3.48
	P	125.2	119.8	111.8	150.1	–0.86	–1.34	17.13	1.66
	Y	1,124.0	1,155.0	1,075.0	1,427.0	0.55	–1.39	16.37	2.25
Mizoram	A	49.0	51.3	65.8	68.1	0.94	5.65	1.75	3.25
	P	45.3	63.8	101.5	110.6	8.17	11.82	4.48	12.01
	Y	924.0	1,244.0	1,543.0	1,624.0	6.93	4.81	2.62	6.31
Nagaland	A	120.8	127.4	140.0	145.0	1.09	1.98	1.79	1.67
	P	130.0	156.3	185.0	187.0	4.05	3.67	0.54	3.65
	Y	1,076.0	1,227.0	1,321.0	1,290.0	2.81	1.53	–1.17	1.66
Tripura	A	280.6	274.0	231.5	257.8	–0.47	–3.10	5.68	–0.68
	P	389.6	501.3	465.5	535.8	5.73	–1.43	7.55	3.13
	Y	1,388.0	1,830.0	2,011.0	2,078.0	6.37	1.98	1.67	4.14
North-East Region	A	3,298.5	3,325.6	3,317.8	3,396.5	0.16	–0.05	1.19	0.25
	P	3,986.8	4,528.1	4,710.1	4,847.6	2.72	0.80	1.46	1.80
	Y	1,209.0	1,361.0	1,420.0	1,427.0	2.51	0.87	0.25	1.50
All India	Y	1,417.0	1,751.0	1,797.0	1,895.0	4.71	0.53	2.73	2.81

Complied from *Basic Statistics of North-Eastern Region, 2000*, NEC, Shillong.

double cropping, especially the winter-rice is showing impressive results.

Table 2.2: Index Number of Rice Yield (All India = 100)

States	1985-86	1990-91	1995-96	1997-98
Arunachal Pradesh	76.78	66.82	58.38	56.94
Assam	81.51	74.99	75.35	71.72
Manipur	142.55	99.49	119.59	117.52
Meghalaya	79.32	65.96	59.82	75.30
Mizoram	65.21	71.05	85.87	85.70
Nagaland	75.94	70.07	73.51	68.07
Tripura	97.95	104.51	111.91	109.66
NE Region	85.32	77.73	79.02	75.30
All India	100.00	100.00	100.00	100.00
1985-86 = 100	100.00	123.57	126.82	133.73

The key to agricultural prosperity lies in the growth of yield, which from Tables 2.1 and 2.2 indicate a dismal picture. Except for the two States of Manipur and Tripura where the rice yield is around the national average (though with some decline in case of Manipur), the average rice yield grew at a rate of 1.5 per cent during the entire period, and was less than 0.5 per cent during the nineties (increased from 1,361 kg. to 1,427 kg. per hectare between 1990-91 to 1997-98). During this period, the All India average rice yield increased from 1,751 kg. to 1,895 kg. per hectare.[1] This is, of course not to compare with the best performing rice States like Punjab, Haryana, Tamil Nadu, Andhra Pradesh, or even the neighbouring West Bengal[2] where the average yield is around 2,500 kg./ha. At the level of the individual States, over the entire period the yield growth has been substantial in two States only, Mizoram (6.31%) and Tripura (4.14%); in rest of the States, either the yield increases have been negative as in Arunachal Pradesh and Nagaland or only marginal, well below the national average. Surprisingly, in most of the States the yield growth in end eighties was reasonably higher than during the nineties, a definite cause for concern. This must be viewed against the agro-climatic backdrop in which unlike many parts of India the monsoon regime is fairly stable over the North-East Region, with little scope for blaming an erratic monsoon. Indeed, as indicated by the index number, not only the figures are well below the 100 marks (except,

Manipur), they are on the decline, further leading one to conclude that by and large the States are falling behind in comparison to the All India average picture.

Land Distribution

Undoubtedly the most important asset that an average farming family in India holds is farm land, which with increasing population pressure, holding fragmentation through successive generations and without any major population shifts to non-agricultural occupations have made these farm holdings unremunerative, and the farmers (cultivators) virtually unemployed even with their own lands. As some studies from prosperous agricultural States like Punjab and Haryana indicate, small and marginal farmers discovered a better option of either selling out the small parcels at a remunerative price or even renting out to rich farming neighbours and migrating out to towns/cities for non-farm wage work. An overall growing economy may provide such non-farm opportunities unlike the stagnant one as in the North-East. Traditionally, North-East Region used to be relatively low on population and without the influence of the notorious zamindari system (except for parts of Goalpara district); by and large the Assamese cultivator had some land often neither too large or too small holdings. In the hills, land question was unimportant till about recent times. All this have changed in the region and are changing very fast to the detriment of small and marginal farmers.

Table 2.3 highlights the State-wise size classes of operational holdings (5 classes) for 1995-96 in percentages of area and number of holdings as well as the Ginni Ratios (Coefficients) indicating the level of inequalities in land assets in the States. As compared to 59 per cent of farmers in India having less than one ha holdings (very small), Assam comes the closest with 60 per cent, Tripura 68 per cent, Mizoram 48 per cent and the least in Nagaland 9 per cent only. On the other hand, in the large holding category (10 ha +), in Nagaland 23 per cent of farmers have 57 per cent of cultivated holdings, as compared to country average of 1.59 per cent having 17.45 per cent of area, in Assam 0.25 per cent having 14 per cent large farms (area). Mizoram has no large holdings as such, so also Tripura and Manipur (negligible). With the exception of Nagaland, the general expectation of revealing an egalitarian asset distribution

Table 2.3: Size of Operational Holdings and Ginni Ratios (1995-96)

Size Classes/ States (in ha.)	Arunachal Pradesh		Assam		Manipur		Meghalaya		Mizoram		Nagaland		Tripura		All India	
	Area	Num.	Area	Num.	Area	Num.	Area	Num.	Area	Num.	Area	Num.	Area	Num.	Area	Num.
0 to 1	2.91	17.89	18.98	59.98	21.71	48.59	10.96	36.42	21.43	47.54	0.93	9.15	28.25	68.24	14.87	58.99
1 to 2	7.85	18.95	24.07	22.57	38.29	34.51	19.86	25.93	42.86	37.70	3.10	14.79	34.42	21.70	17.34	18.97
2 to 5	24.71	31.58	27.65	13.39	31.43	14.79	40.07	28.40	29.76	14.75	7.85	18.31	25.00	8.81	23.16	13.21
5 to 10	43.02	27.37	15.22	3.80	8.57	2.11	24.32	8.02	4.76	1.64	30.79	33.10	5.84	1.26	27.20	7.25
10+	21.51	4.21	14.08	0.25	0.57	0	4.45	0.62	0	0	57.44	23.24	6.49	0	17.45	1.59
Ginni Coefficients	0.44		0.53		0.35		0.41	0.34		0.43		0.45		0.58		

Source: CMIE, 1999.

is belied in the region. The Ginni ratios in Assam follows closely the pan Indian pattern, only in three smaller States of Mizoram and Manipur that one encounter Ginni ratios somewhat lower (0.34 and 0.35 respectively). Without a comparable series in recent times, it is difficult to comment on the process; the quinquennial agricultural censuses have become erratic and the publications, even more unpredictable. However, the impression one gets is that the landholdings inequalities in the region are on the rise, so also landlessness, especially in the hilly tracts where the earlier traditions of egalitarian land distribution has gone with the wind, especially that the conversion of community held land in rural areas have been constantly and swiftly replaced by private property. The consequences are not far to seek.

Rural Unemployment

Since 85 per cent of the population in North-East live in rural areas (87% in case of Assam) expectedly employment essentially means rural and necessarily agricultural. If agriculture is growing, i.e., there is some dynamism in it, then with additional disposable incomes with farmers, one would expect buoyancy in both farm and non-farm work in rural areas.

Table 2.4 indicates that whereas the overall unemployment level from 1983 to 2000 has remained more or less constant (at about 2%), in many North-Eastern States it is on the rise, e.g., Nagaland from 0.3 to 3.5 per cent (10 times), Manipur from 0.4 to 3.5 per cent, Assam from 2.2 to 4.6 per cent and for reasons cited above, these could all arise out of rural unemployment. The disconcerting feature appears to be a steady rise in female unemployment, e.g., Assam 8 per cent, Manipur 3.1 per cent, Tripura 4.5 per cent.

Rural unemployment was estimated by Srivastav and Dubey (2004)[3] for 1993-94 (50th round NSSO), which indicates 7.64 per cent for Assam (female 14.56%), North-East total, 5.42 per cent (female 6.94%) as compared to the All India average of 1.97 per cent (female 1.50%), also because of the large weightage of Assam in North-East population. This indicates quite a disturbing situation: (a) the rural unemployment in the region is 3 times the All India average, and (b) female rural unemployment nearly 5 times the All India average. There is undoubtedly some serious issues involved in

Table 2.4: Unemployment in North-Eastern States

(in per cent)

States	1983			1993-94			1990-2000		
	Male	Female	Total	Male	Female	Total	Male	Female	Total
Arunachal Pradesh	—	—	—	1.3	0.5	1.0	1.0	0.9	0.9
Assam	2.2	2.4	2.2	4.6	9.5	5.6	3.7	8.0	4.6
Manipur	0.6	0	0.4	2.2	1.0	1.8	3.7	3.1	3.5
Meghalaya	1.8	0.9	1.5	0.5	0.3	0.5	0.9	0.9	0.9
Mizoram	0.3	1.0	0.4	0.9	0.5	0.8	2.5	1.3	2.0
Nagaland	0.4	0	0.3	3.2	0.6	2.4	4.0	2.9	3.5
Tripura	2.1	17.7	3.9	2.2	8.4	3.4	1.6	4.5	1.9
All India	2.3	1.9	2.0	2.1	1.7	2.0	2.5	1.8	2.3

Source: P.C., NHDR, 2001 (Based on NSSO, 38th, 50th and 55th rounds).
Note: No samples from Arunachal Pradesh for 1983.

the agrarian sector in the region—even the much maligned Bihar is better off. What good is agricultural development, if any, if the villagers have no work? Second, all egalitarian theorisation of better gender situation, literacy, etc. ascribed to the region falls flat on the face of huge rural female unemployment, many times the rest of India.

Rural Poverty

The expectation that a growing agriculture could bring down the level of rural poverty in never automatic with out appropriate enabling public policies, like food subsidies to the poor, crop insurance and other social sector inputs. Indeed, experience of some developed agricultural States like Punjab, Andhra Pradesh and Karnataka, etc. indicate deteriorating condition of small and marginal farmers including land alienation, suicides (due increased debt burden and crop failures). On the other hand, subsidies on fertiliser, water and electricity have a built-in favourable bias to the rich farmers. These issues are well known. We are in any case, not very clear about higher agricultural growth and reduction in rural poverty, with possible exception of the State of Punjab, where urban poverty exceeds rural poverty.

Table 2.5 indicates two different sets of HCRs for all the seven States of the region (and for comparison, India average) for 1987-88, 1993-94 and 1999-2000. The first set refers to Bandopadhyay and Dubey (relevant here in Adjusted Poverty Line) for 1987-88 and 1993-94 and then the Planning Commission figures (NHDR, 2001) for 1993-94 and 1999-2000. At the All India level (1993-94) the estimate differs by about 5 per cent (APL: 30.46 to PC: 35.97%). At the All India level the overall rural poverty reduction between 1987-88 and 1993-94 has been by 6.5 per cent and between 1993-94 and

Table 2.5: Rural Poverty in North-Eastern States (1987-88, 1993-94 and 1999-2000)

States	1987-88			1993-94			1993-94	1999-00
	OPL	EOPL	APL	OPL	EOPL	APL	HCR	(PC*)
Arunachal	30.14	26.43	19.08	48.75	44.66	34.08	39.35	33.47
Assam	48.42	39.75	32.05	57.05	48.00	40.47	40.86	36.09
Manipur	21.11	15.82	9.52	33.08	23.91	13.74	33.78	28.54
Meghalaya	40.39	36.18	29.74	34.36	28.36	18.26	37.92	33.87
Mizoram	5.50	3.93	2.76	10.10	7.05	4.11	25.66	19.47
Nagaland	—	—	—	4.24	3.18	1.91	37.92	32.67
Tripura	28.91	23.73	18.37	32.04	26.13	21.21	39.01	34.44
All India	49.61	39.76	36.91	42.82	33.41	30.46	35.97	26.10

Source: The first set of figures (1987-88 and 1993-94) are from, Gangopadhyay & Dubey, 2000 and the second set (1993-94 and 1999-2000) from NHDR, PC, 2001.

1999-2000 by 9.9 per cent (35.97 to 26.10%). How do the North-Eastern States compare? In Arunachal, the first period had an increase by 15 per cent and second, a reduction by 6 per cent—net effect, negligible. In Assam, during the first period there was an increase by 8.5 per cent; second, reduction by 4.5 per cent—net effect, it has a rural poverty level 10 per cent higher than All India average. In Manipur, the first period was witness to an increase by 4 per cent and later period, a decline by 5 per cent—net effect negligible. In Meghalaya, during the first period there was a decline by 11 per cent, but later on, only 4 per cent—net effect, the rural poverty ratio still way above the All India average. It is only in Mizoram that one notices the rural poverty levels way below the national average, but that was the case to start with. It may be noted that during 1981-91,

Mizoram was the only State with a net absolute decline in rural population (urbanisation level jumped from 24 to 46%), which indicate a strong stream of rural to urban migration as well as declaration of larger settlements urban (in 2001, it is 49.5%). In Nagaland, from earlier very low level, the rural poverty by recent estimates (PC) has reached 39 per cent, which decline in 1999-2000 to 33 per cent, by 6 percentage points still way above the national average. In Tripura, it increased by 8 per cent during the first period, but declined by 5 per cent in the latter period, still way above the national average. What does one make out of it? Except for Mizoram, all the other States have rural poverty levels above the national average, even by different estimates. Agriculture, that is the main stay of these predominantly rural States definitely has not delivered. There is the tail-tell signal of something wrong in the basics of the situation.

Conclusion

The four tests we carried out all bore wrong signatures:

(1) Since, rice farming is the main stay of agriculture, no apparent improvement (from mid-80s to end-90s) in land productivity (except, Mizoram) indicates lack of dynamism in the rural economy—the trend is not progressive on the wealth generation end in the rural economy.
(2) The land distribution has perhaps worsened. But in any case, contrary to belief the inequality in land asset distribution is nothing better than the rest of the country.
(3) Undoubtedly, the rural unemployment has only worsened over the decade. North-East as a whole has on an average 3 times more rural unemployed as compared to All India average.
(4) The poverty level in rural areas (except Mizoram) is appreciably higher than All India average and the decline if any has been much slower as compared to the national figures.

The story of agricultural changes appears to be a dismal one for the region. But more important, the consequences of a sluggish agriculture on rural unemployment and poverty are significant. How this could be reversed, is the moot question.

NOTES

1. The overall rice yield growth over the 12-year period has been 2.81 per cent per year for India, well above the population growth.
2. West Bengal's agricultural performance has been sterling in the past decade.
3. Nirankar Srivastav and A. Dubey, "Unemployment in North-East India: Some Issues", A.C. Mohapatra and C.R. Pathak (ed.), *Economic Liberalization and Regional Disparities in India*, Star Publication, Shillong, pp. 202-220.

REFERENCES

Bandopadhyay, S. and Dubey, A.C. (1999). *Counting the Poor*, CSO, New Delhi.

Bharadwaj, K. (1975). *Production Conditions in Indian Agriculture*, Cambridge UP, Cambridge.

CMIE, (1999). *Agriculture*, Mumbai.

Dreze, J and Sen, A.K. (1996). *Indian Development*, OUP, Delhi.

E.H. Carr (1965). *What is History?*, Penguin, Harmondsworth, Middlesex, pp. 56-86.

NEC (2000). *Basic Statistics*, Shillong.

Planning Commission (2002). *National Human Development Report*, OUP, New Delhi.

Srivastav, Nirankar and A.C. Dubey, "Unemployment in North-East India: Some Issues", A.C. Mohapatra and C.R. Pathak (ed.), *Economic Liberalization and Regional Disparities in India*, Star Publications, Shillong, pp. 202-220.

3

RETHINKING OF AGRICULTURAL DEVELOPMENT NEED OF NORTH-EAST FARMERS

— D.N. CHAKRAVARTY

Small and peasant farmers are the primary producer of staple foods, accounting for very high percentage of national production in our country. But this sector in itself is characterised by poverty and hunger and in so many cases lagging productivity. The history of North-East Region since the beginning has been the history of subsistence farming. The colonial land grabing and affluents particularly in Assam pushed rural food producing societies from the best lands most suitable for farming.

The farming people accustomed to the continuous production of annual crop in fertile lands with good access to water were driven to marginal areas. As a result forests were felled and many fragile habitats were subject to unsustainable production practices. These marginal lands are characterised by broken terrain, slopes, high and irregular rainfall, little or no irrigation and/or low fertility. This resulted in producing newly destitute and displaced farmers. This dynamics drives environmental degradation to an alarming level.

Moreover, being poor they are victimised by pervasive anti-poor and anti-small farmer biases. As such their agriculture is best characterised as complex, diverse and risk-prone.

Little effort has been made to alleviate these environmental and social problems generated by this dynamic which ushered in a new era of land concentration in the hands of the wealthy, and drove the growing problem of landlessness in rural areas. The landless rapidly became the poorest of the poor, subsisting as part-time seasonal agricultural or day labourers, share croppers or migrating to the agricultural frontier to fell forests for homesteads.

Thus the rural areas of North-East are characterised by extreme inequalities in access to land, insecurity of land tenure and in the quality of land farmed. By keeping wages and living standard low, the elite guarantee that healthy domestic markets will never emerge. This will reinforce export orientation. The result is a downward spiral into deeper poverty and marginalisation.

One irony of our world is that food and other farm products flow from areas of hunger and need to areas where money is concentrated.

The situation is no better in the more favourable lands. In these soils more and more fertilisers and pesticides are being used in some localised areas on North-East Region and thereby exposing these soils to the hazards of chemical agriculture. Though these hazards are alarming in the States such as Punjab and Haryana the situation in this part of the country is not yet deteriorated to that extent. If high level of chemical fertilisers and pesticides continued to be used the productive capacity of these soils also will drop due to pollution, erosion, waterlogging and fertility loss, together with growing resistance of pests to pesticides and the loss in soil and functional bio-diversity.

On the other hand, the agricultural, social and infrastructural development programmes to ensure social justice of poor and vulnerable people have not gained actual momentum. The majority of the poor still live in rural areas, and though these changes could drive them to some extent many of them plunge into new depth of crisis in sustaining their livelihood. Small farmers find the prices of staple foods they produce dropping below the cost of production. The lagging of the productivity of the soil of the poor people is due to displacement of the poor to marginal, rainfed lands, and face structures and macro-economic policies that are increasingly unfavourable to food production by marginal farmers. Under these circumstances no new technology or new variety of seeds, fertilisers and pesticides can change the situation without simultaneous structural changes in access to land and agricultural and trade policies.

Such farmers have rarely benefited from "top down" formal institutional research and Green Revolution technologies. Any new strategy truly addressed to productivity and poverty concerns will have to meet their needs for multiple suitable crops and management practices. Formal research methods are not able to handle the vast complexity of physical and socio-economic conditions of these small and marginal farmers.

In order to survive under such circumstances and improve their standard of living we must be able to tailor agricultural technologies to their variable but unique situation, in terms of local climate, topography, soil, biodiversity, cropping systems, market facilities, available resources, etc. For this reason farmers over the ages are involved in complex farming and livelihood systems which balance risks. Their cropping systems typically involve multiple annual or perennial crops, animal, poultry, fish and variety of vegetables and medicinal plants.

The inescapable conclusion is that a different approach is essential by organising the farmers themselves which take into account the multiple characteristics of both agricultural technologies and farmers themselves

Here two major approaches make the most sense under such conditions: (1) technologies which have pro-poor diseconomies of scale like eco-technologies, and (2) organisation into social movements capable of exerting sufficient political pressure to reverse policy biases.

So it is the time that requires rethinking and rediscovering our technological base for future plan of work for sustainable agricultural development. As we approach a new century we can look back and draw a balance sheet of our achievements and failures. In this age of rapid changes, transition in many respects to the renaissance is triggered by rapid pace of technological changes, information revolution, globalisation and interaction of technology and society. Centrifugal forces of globalisation are blowing away some established notions and format while forging new links. The country has to shift to the new era in order to get benefit from opportunities and challenges and learn to turn the crisis into advantages.

To cope with the present crisis in agriculture the available set of technologies are not sufficient and suitable particularly for small farm level activities. Rural people do not take risk to adopt a new technology immediately. So every new technology finds a resistance at the beginning. The need of the hour for the scientific community, academic institutions and society as a whole is to direct their energies towards reducing dependence on polluting technologies, non-relevant methods and products and generate relevant, easily degradable eco-friendly products and systems.

The technologies for sustainable agricultural development is rooted in the principles of ecology, economics and equality and is popularly known as ecotechnology. UNESCO and the Cousteau Foundation are promoting ecotechnology network in different parts of the world. The M.S. Swaminathan Research Foundation at Chennai is the co-ordinating centre for Asia. The basic principles of ecotechnology (M.S. Swaminathan, *The Hindu*, Novenber 3, 1996) aims at production of food from efficient and environmentally benign production technologies that conserve and enhance the natural resources base of crops, animal husbandry, forestry and fisheries.

The productivity of rainfed rice in North-East Region of India is constrained by the unfavourable biophysical and complex socio-economic conditions. The farmers of this region particularly in the State of Assam have to operate their rice cultivation practices under stressful situation such as unpredictable and erratic rainfall resulting in flood, waterlogging, drought, nutrient deficiencies and toxicities, weeds, insect pests and diseases pressures. This led to farming situation of heterogeneous and variable nature. This situation emphasise the need for relevant technologies as per climate variation, soil characteristics and position in the landscape, depth of water and other variables. The technologies already developed should be perfected as per agro-ecological situation, economic conditions and the perception and aspiration of the farmers.

For hill areas special attention is needed to collaborate technological development with the type of conservation methods adopted, slope and exposer of the hills, moisture regime, etc. over and above the socio-economic aspects. To achieve this a self-propelling and self-replicating system is to be devised so that the programme does not collapse when external helps and subsidies are withdrawn.

Along with ecotechnology training modules should be developed to enable villagers to prepare socio-demographic character of their respective villages, turn information techniques to established information centres where information in the meteorological parameters, management techniques and marketing factors relevant to rural livelihood will be converted into location specific information.

While new information techniques are important folk media are often even more effective in reaching the target group. The basic approach, therefore, revolves around imparting a pro-nature, pro-poor, pro-women, pro-youth orientation to a job-led economic growth strategy. There should also be an inventory of opportunities with relevant technologies open for women and promote greater involvement of women both in technology development and transfer.

Biology and biotechnology would play a crucial role in enhancing the quality and quantity of foodgrains like rice, wheat and other crops, healthcare, employment, safe environment and economic opportunities. Tissue culture technique at cellular level and genetic engineering at gene level have provided new opportunities for obtaining genetically superior high quality planting material with desirable traits. Vermiculture is not a new technique and the age old process was known to farmers as the earthworms burrowing around and transforming organic matter into rich manure.

Indigenous knowledge is not only the foundation of modern science, it is also the base of the references and referral centre for modern plant breeding. These indigenous knowledge and expertise should be explored and blended with modern technologies for developing the future technologies, for improving production of rice and other foodgrains for fighting the poverty in real sense.

Participatory research with farm family's help and dissemination of the results in the form of ecotechnologies has already gained momentum in many States of the country. Integrated pest management and integrated nutrient supply are some of the programmes where farmers can participate. For the purpose of appropriate ecotechnology development and dissemination there is need of grouping farmers as per their economic condition and capacity to adopt a technology, for example resource rich farmers (to provide them with technical guidance only), Green Revolution farmers (who consumed the produce and sold the rest) and poor farmers (majority of whom are illiterate and should get total financial and technological support).

Science and technology are not culture free. A society or a community accept a technology through a variety of mechanism and in various stages, through a process of evaluation that is still not very clearly understood. That is why there is a need to combine technologies with the understanding of human ways and needs. New technologies create new cultures.

Our society's rich cultural activities are associated with rice culture. Biodiversity in rice culture is linked with cultural diversity of the region. In a book, *Our World's Heritage*, published by National Geographic Society (1987), 322 numbers of well-linked cultural and natural diversities are recorded. We must start our study for development of technology from work engagement of the people in the society, which is the basic for development. It is actually an attractive idea of understanding the present in terms of the past. It is worth examining the continuity between tradition, myth and reality. A better understanding of the power of tradition and its hold in the society's collective consciousness is necessary.

Development of agricultural technology is only a sub-system of overall development. Both food and non-food items assume importance in determining the nutritional security of a society. While we succeed in the diligent development of technological and productivity potentials we could lose all the advantages for backlog or disadvantages in other sub-systems such as lack of inputs, unavailable water and power supply, inadequate communication facility and transportation and lack of other facilities. The scientific community need to voice their concerns in public affairs with the administration and play a more significant role in ensuring an integrated development of the economic progress and quality of life of the rural people.

4

STATUS OF AGRICULTURAL SECTOR IN NORTH-EAST INDIA
Problems and Prospects

— C. HAZARIKA

The North-Eastern Region reflects ecological and cultural contrast between the hills and the plains and yet there are also significant elements of continuity. The North-East is one of the most biodiverse regions in the world. This chapter attempts to study the agricultural scenario of the North-Eastern Region and to identify the problems and prospects of agricultural sector. Secondary information published by various sources were used for the present study.

Agriculture

The economy of the North-Eastern States is mainly rural and agrarian. The region offers scope for cultivation of a wide variety of agricultural crops because of its diversities in topography, altitude and climatic conditions. These products are mainly sold in the local markets, and mostly in the form of primary produce without significant value addition.

Land to Man Ratio

Land to man ratio is an important parameter of development of agricultural sector. The extent of cultivable land in the North-Eastern Region varies from State to State. Land is a critical resource in many of the North-East States and availability and management of land for agricultural activities are essential for raising the region's overall agricultural production and productivity. Table 4.1 presents the geographical land to man ratio of the different North-Eastern States as on 1991. The region's agricultural system is predominantly traditional. The overall geographical land to man ratio for the North-Eastern Region (0.81 ha/person) is much higher than the national average (0.39 ha/person). Among the North-Eastern States the geographical land to man ratio is highest in Arunachal Pradesh (9.69 ha/person) and lowest in Assam (0.35 ha/person).

Table 4.1: Geographical Land to Man Ratio (Hectares/Person) as on 1991

State	Geographical area (Ha)	Population (No.)	Ratio
Arunachal Pradesh	83,74,300	8,64,558	9.69
Assam	78,43,800	2,24,14,322	0.35
Manipur	22,32,700	18,37,149	1.22
Meghalaya	22,42,900	17,74,778	1.26
Mizoram	21,08,100	6,89,756	3.06
Nagaland	16,57,900	12,09,546	1.37
Tripura	10,48,600	27,57,205	0.38
All North-Eastern States	2,55,08,300	3,15,47,314	0.81
All India	32,87,26,300	84,63,02,688	0.39

Size of Holding

Size of holding is another important variable for agricultural sector development. About 80.00 per cent of the farmers in the North-Eastern Region belongs to the small (< 1.44 ha) and marginal (< 0.40 ha) category. Moreover, with the increase in population, the average size of holding is gradually reducing over the years. Table 4.2 presents the average size of holding in North-Eastern States during 1990-91. The average size of holding for the North-Eastern States (1.60 ha) is higher than the all India (1.57 ha) average. Among the North-Eastern States the average size of land holding is highest in Nagaland (6.92 ha) and lowest in Tripura (0.97 ha). About 55.60 per cent of the holdings are of size only 0.43 ha or less. This is primarily because hilly terrain constitutes nearly two-thirds of the region's geographical area, and large sized holdings are not feasible. The average plot size is too small for mechanisation of agriculture and adoption of modern farming practices. Subsistence farming, therefore, is predominant in the North-Eastern Region and there is hardly any commercial surplus.

Total operational land in the North-Eastern Region is 53.4 lakh ha which is 3.22 per cent of total operational land in the country. The highest operated area of 31.6 lakh ha is in Assam and the lowest (0.8 lakh ha) is in Mizoram. Nagaland has the unique characteristics of highest size of operational holding at 6.92 ha, which is much higher than the average holding size of the region (1.60 ha) and the average holding size of all India (1.57 ha).

Table 4.2: Average Size of Holding (in Hectares) 1990-91

State	Marginal	Small	Semi-medium	Medium	Large	Average holding size
Arunachal Pradesh	0.60	1.50	2.81	5.76	18.43	3.62
Assam	0.41	1.39	2.70	5.23	74.17	1.31
Manipur	0.55	1.37	2.56	5.01	12.16	1.24
Meghalaya	0.54	1.38	2.54	5.46	13.00	1.81
Mizoram	0.62	1.57	2.78	4.00	0.00	1.34
Nagaland	0.69	1.43	2.92	6.34	16.85	6.92
Tripura	0.40	1.53	2.69	5.14	121.57	0.97
All North-Eastern States	0.43	1.41	2.70	5.56	25.20	1.60
All India	0.40	1.44	2.76	5.90	17.30	1.57

Cultivation Practices in North-Eastern States

Table 4.3 presents the cultivation practices in the North-Eastern States during 1995-96. The percentage of cultivable area over the geographical area in most of the North-Eastern States (24.69%) is less than the national average (59.22%). The percentage of cultivable area over the geographical area is highest in Meghalaya (47.88%) and lowest in Arunachal Pradesh (3.60%). The percentage utilisation of cultivable area in the North-Eastern region (62.04%) is less than the national average (73.05%) and mono cropping is the predominant method of cultivation. In the absence of multi-cropping, little or no surplus is generated in the agricultural sector. The utilisation is lowest in Meghalaya (19.18%) and highest in Tripura (89.35%). Among all the North-Eastern States, Assam takes the highest position both in terms of available cultivable area and net sown area.

Table 4.3: Cultivation Practices in North-Eastern States during 1995-96

State	Total geo-graphical area ('000 ha)	Cultivable land ('000 ha)	% of geo-graphic area	Total crop area ('000 ha)	Area sown more than once ('000 ha)	Net sown area ('000 ha)	% of cultivable area used	Gross irrigated area ('000 ha)	% of gross cropped area
Arunachal Pradesh	8374.3	293	3.50	244	59	185	63.14	36	14.75
Assam	7843.8	3387	43.18	3938	1158	2780	82.08	572	14.53
Manipur	2232.7	164	7.35	182	42	140	85.37	75	41.21
Meghalaya	2242.9	1074	47.88	247	41	206	19.18	45	18.22
Mizoram	2108.1	445	21.11	109	0	109	24.49	9	8.26
Nagaland	1657.9	626	37.76	228	17	211	33.71	72	31.58
Tripura	1048.6	310	29.56	426	149	277	89.35	60	14.08
All NE States	25508.3	6299	24.69	5374	1466	3908	62.04	869	16.17
All India	328726.3	194680	59.22	186561	44346	142215	73.05	71510	38.33

Source: CMIE Report on Agriculture, 1997 and 1998.

Progress of Agriculture

Table 4.4 presents the progress of agriculture in terms of foodgrains in the North-Eastern States during the period from 1986-87 to 1996-97. Almost 90.00 per cent of the region's rural population is entirely dependent on agriculture. In the period of 1986-87 to 1991-92, for the North-Eastern States the CARG of agricultural area

Table 4.4: Progress of Agriculture (Foodgrains) in the North-Eastern States

State	Factors	1986-87	1991-92	*CARG	1991-92	1996-97	*CARG
Arunachal	A	175.0	188.7	1.52	188.7	186.8	–0.20
Pradesh	P	187.0	219.8	3.28	219.8	219.6	–0.02
	Y	1069	1165	1.73	1165	1176	0.19
Assam	A	2579.8	2749.9	1.29	2749.9	2728.7	–0.15
	P	2588.0	3379.2	5.48	3379.2	3532.1	0.89
	Y	1003	1229	4.15	1229	1294	1.04
Manipur	A	173.0	164.3	–1.03	164.3	172.5	0.98
	P	258.0	350.1	6.30	350.1	390.7	2.22
	Y	1491	2131	7.40	2131	2265	1.23
Meghalaya	A	138.8	133.5	–0.78	133.5	133.1	–0.06
	P	131.3	155.3	3.41	155.3	151.9	–0.44
	Y	946	1163	4.22	1163	1141	–0.38
Mizoram	A	52.7	66.0	4.60	66.0	77.5	3.26
	P	52.8	90.2	11.30	90.2	133.8	8.21
	Y	1002	1367	6.41	1367	1726	4.77
Nagaland	A	147.0	182.4	4.4	182.4	199.5	1.81
	P	93.0	203.6	16.97	203.6	211.9	0.80
	Y	633	1116	12.01	1116	1062	–0.99
Tripura	A	286.0	272.5	–0.96	272.5	272.7	0.01
	P	374.0	486.7	5.41	486.7	556.1	2.70
	Y	1308	1786	6.43	1786	2039	2.69
North-East	A	3552.3	3757.3	1.13	3757.3	3770.8	0.07
	P	3684.1	4664.9	5.80	4664.9	5196.1	1.24
	Y	1037	1300	4.62	1300	1378	1.17
All India	A	127197.0	121871.2	–0.85	121871.2	124509.5	0.43
	P	143419.0	168373.2	3.26	168373.2	199321.2	3.43
	Y	1128	1382	4.15	1382	1601	2.99

Note: * CARG = Compound Annual Rate of Growth.
A = Area ('000 ha), P = Production ('000 tonnes), Y = Yield (kg/ha)

Source: CMIE Report on Agriculture, 1997 and 1998.

(1.13%), foodgrain production (5.80%) and foodgrain yield (4.62%) were higher than the national average of the CARG of agricultural area (–0.85%), production (3.26%) and yield (4.15%). But significantly in the period of 1991-92 to 1996-97, for the North-Eastern States the CARG of agricultural area (0.07%), production (1.24%) and yield (1.17%) have come down compared to the national average of CARG of area (0.43%), production (3.43%) and yield (2.99%). In the period of 1986-87 to 1991-92, the performance of North-Eastern agricultural sector was quite good which could compete with the Indian agriculture, but in the period of 1991-92 to 1996-97, the North-Eastern agriculture could not maintain the pace of improvements of agriculture with the Indian agricultural scenario. This might be due to low usage of HYV seeds, fertilisers and other agricultural inputs like irrigation, credit, improved agricultural implements and use of primitive and outdated techniques.

Per Capita Availability of Foodgrains

The per capita availability of foodgrains (Table 4.5) in Arunachal Pradesh (254 kg/annum) is the highest among the North-Eastern States (164.71 kg/annum) and even higher than the national average (235.52 kg/annum). The per capita availability of foodgrains is lower than the national average in all the other North-Eastern States.

Table 4.5: Per Capita Availability of Foodgrains in the North-Eastern States Based on the 1991 Population Census (kg/annum)

State	Per capita availability of foodgrains (kg/annum)	
	1991-92	1996-97
Arunachal Pradesh	254.23	254.00
Assam	150.76	157.58
Manipur	190.57	212.67
Meghalaya	87.50	85.59
Mizoram	130.77	193.98
Nagaland	168.33	175.19
Tripura	176.52	201.69
All NE States	154.84	164.71
All India	198.95	235.52

Shifting Cultivation

Shifting cultivation (*jhum* cultivation) is the common practice of agricultural production in the hills. Recently terrace cultivation is encouraged through State government initiatives. The annual area under shifting cultivation is 3,809 sq. km. and minimum area under shifting cultivation in one time is 14,660 sq. km. and 4.43 lakh families are practicing it (Table 4.6). The *jhum* cycle varied from 3-10 years but due to pressure of population the same comes down to 2-3 years. Shifting cultivation has its own danger of soil erosion, soil degradation, loss of valuable forest products and flooding in river valley down below. All these cause ecological imbalance. Settled cultivation is encouraged through various programmes like:

1. Permanent settlement of *jhumia* cultivators through development of plantation crops in Karbi Anglong and NC hills of Assam.
2. Providing 2.0 ha of terrace land to *jhumia* family along with inputs and financial help for permanent cultivation in Meghalaya.
3. Pilot project in selected villages, land reclamation, minor irrigation, land improvement, provision of seeds, fertilisers and development of horticulture and cash crops in Mizoram.
4. Pilot projects in Nagaland to induce farmers to give up *jhum* cultivation and adopt terrace cultivation.

Table 4.6: Shifting Cultivation in North-East Region

State	Annual area under shifting cultivation (sq km)	Fallow period (years)	Minimum area under shifting cultivation one time or other (sq km)	No. of families practising shifting cultivation
Arunachal Pradesh	700	3–10	2100	54000
Assam	696	2–10	1392	58000
Manipur	900	4–7	3600	7000
Meghalaya	530	5–7	2650	52290
Nagaland	190	5–8	1913	116046
Tripura	223	5–9	1115	43000
All NE States	3809 (1.5%)		14660 (5.7%)	443336

Source: Basic Statistics, NEC.

Changes in Cropping Pattern

Rice is the main crop in the North-East Region covering around 61 per cent of gross cropped area. Maize is the next important crop for all the hill States except Tripura and Assam. Oilseed is another important crop. During the period 1970-71 and 1995-96 changes in cropping pattern was observed (Table 4.7). This period is marked by onset of Green Revolution in seventies, commercialisation in agriculture and tremendous increase in production due to adoption of improved technology and yield increasing strategies, introduction of cash crops and intensive cultivation practices. In the North-East Region the main changes were as follows:

1. Rice continued to remain as the major crop in all the States. However, the proportionate area declined in Assam and Tripura, remained stable in Manipur but increase in Mizoram and Meghalaya.
2. Oilseeds are the next important crop covering around 6-8 per cent of area except Manipur, Meghalaya and Tripura. Rape and mustard was the only oilseed crop till seventies but after that new oilseed crop like linseed, castor, sunflower, groundnut has entered into the cropping pattern.
3. Wheat is grown mainly in Assam with tremendous area expansion in the late seventies due to Green Revolution. But since 1985-86, wheat area is declining due to lack of irrigation facilities, problem of pre-harvest sprouting in the event of early showers and problem of post-harvest processing.
4. Pulse area is remaining stagnant in Assam due to problem of soil moisture stress in the sowing season, poor management, disease and pest, etc. this crop is newly entered in Arunachal Pradesh, Mizoram, Nagaland and Tripura in eighties.
5. Jute and sugarcane are the two important cash crops of the region. Due to problem of retting, processing and marketing the jute area is declining. Rice is substituting jute area.
6. Potato is gaining priority in all the States except Mizoram.

Table 4.7: Changes in Cropping Pattern in North-East States

(*Area in lakh ha*)

Crops	Period	Arunachal Pradesh	Assam	Manipur	Nagaland	Tripura	Mizoram	Meghalaya
(1)	(2)	(3)	(4)	(5)	(6)	(7)	(8)	(9)
Rice	1970-71	0.60 (46.15)	21.37 (73.97)	1.50 (80.21)	0.60 (58.82)	2.65 (74.44)	0.05 (6.41)	0.34 (17.08)
	1995-96	1.18 (47.2)	24.93 (62.63)	1.52 (81.28)	1.37 (63.13)	2.48 (53.91)	0.65 (60.75)	1.03 (43.09)
Wheat	1970-71	—	0.13 (0.45)	—	—	—	—	—
	1995-96	0.05 (2.00)	0.82 (2.06)	—	—	0.03 (0.65)	—	0.04 (1.67)
Maize	1970-71	—	0.22 (0.76)	0.06 (3.21)	—	—	—	0.05 (2.51)
	1995-96	0.35 (13.56)	0.19 (0.48)	0.03 (1.82)	0.27 (12.44)	0.02 (0.43)	0.08 (11.76)	0.17 (7.11)
Total pulses	1970-71	—	0.90 (3.11)	—	—	0.03 (0.84)	—	0.02 (1.00)
	1995-96	0.05 (2.0)	1.08 (2.71)	—	0.11 (5.07)	0.10 (2.17)	0.05 (4.67)	0.03 (1.25)
Total oilseeds	1970-71	—	1.73 (5.99)	—	0.02 (1.96)	0.05 (1.40)	0.02 (2.56)	0.05 (2.51)
	1995-96	0.22 (8.8)	3.15 (7.91)	0.04 (2.42)	0.19 (8.75)	0.15 (3.26)	0.07 (6.54)	0.09 (3.76)

(*Contd.*)

Table 4.7: (*Contd.*)

(1)	(2)	(3)	(4)	(5)	(6)	(7)	(8)	(9)
Jute	1970-71	—	1.34	—	—	0.08	—	0.07
			(4.64)			(2.25)		(3.52)
	1995-96	—	0.85	—	—	0.01	—	0.04
			(2.13)			(0.22)		(1.67)
Sugarcane	1970-71	—	0.34	—	0.01	0.02	—	0.002
			(1.17)		(0.98)	(0.56)		(0.10)
	1995-96	—	0.36	0.01	0.04	0.02	0.01	0.001
			(0.90)	(0.60)	(1.84)	(0.43)	(0.93)	(0.04)
Potato	1970-71	—	0.27	—	—	0.03	0.01	0.16
			(0.93)			(0.84)	(1.28)	(8.04)
	1995-96	0.05	0.68	0.04	0.02	0.04	0.005	0.18
		(1.93)	(1.70)	(2.42)	(0.92)	(0.87)	(0.47)	(7.53)
Cotton	1970-71	—	0.04	—	—	0.02	0.01	—
			(0.14)			(0.56)	(1.28)	
	1995-96	—	0.02	0.001	—	0.02	0.01	0.08
			(0.05)	(0.06)		(0.43)	(0.93)	(3.35)

Strategies for Agricultural Development

1. Intensive farming with yield improvement strategies, crop diversification, development of cash crops and high valued crops.
2. Extensive cultivation of prospective crops—expansion of area under summer rice, oilseeds, pulses, horticultural crops, development of medicinal and aromatic plants.
3. Infrastructure development—storage, processing, marketing, irrigation, rural development and communication, input delivery and credit supply.
4. Technology development for different agro-ecological situations—upland, hill areas, and flood effected areas.
5. Strengthening training and demonstration, on farm trials.

Above all, there is an urgent need for investment in agricultural sector in the North-Eastern Region of India. For this purpose, the main agenda should be to attract new investors to this region. In this context, it is imperative to popularise the avenues and opportunities for investment in North-Eastern Region.

Why Invest in North-East India?

- Rich mineral resources.
- Plentiful energy sources—Brahmaputra and its tributaries, forming the largest perennial water system in India, has immense potential for energy, irrigation and transportation.
- Fertile stretches of land.
- Storehouse of horticultural products/plantation crops/ vegetables/spices and rare forest products.
- Abundant forest resources.
- Diverse tourist attractions.
- Reasonably priced and easily available labour.
- Growth centres, total tax free zones.
- Subsidies on transport, capital investment, interest on working capital are available for industries in North-East region.
- Close to Myanmar, the gateway to the ASEAN countries.

- South-East Asia is one of the fastest growing potential markets.
- Proximity to SAARC countries.

North-East India's locational advantage and rich natural resources provide a backdrop to its development as a base for foreign investors.

5

TRANSITION IN AGRICULTURAL GROWTH OF NORTH-EASTERN HILL REGION OF INDIA

— B.K. SARMA, S.B. SINGH and P.P. PAL

'Green Revolution' in India created rich islands in an ocean of poverty leading to many regional and socio-economic imbalances. The new technology was not uniform among different regions, which spread thoroughly in a few favourably endowed States while mountainous region like North-Eastern Hill (NEH) region comprising the States of Arunachal Pradesh, Manipur, Meghalaya, Mizoram, Nagaland, Sikkim and Tripura had lagged behind. This difference in adoption rate of modern agricultural technology may not be entirely due to any policy neglect but could have arisen out of inherent differences in resource endowments and in the levels of infrastructure developments. An analysis of transition in agricultural growth of NEH States reveals that there has been substantial change in the region over the years. But there is a need to examine the extent of regional variation in agricultural growth and the factors associated with it to ensure overall growth in the agricultural economy of the region.

Overall Growth

In the process of economic development, it is expected that non-agricultural sector would grow faster in terms of output and labour absorption. This process was, of course, slow throughout the country (Table 5.1). However, the share of agriculture in net domestic product (NDP) is higher in NEH States except Nagaland, which may perhaps due to comparatively lower industrial growth in the region. The decline in the share of agriculture in NDP is higher than the decline in labour force in agriculture. This coupled with population growth has led to an increase in the number of workers per hectare of land in the NEH region.

Table 5.1: Share (%) of Agriculture in Net Domestic Product and Employment in NEH States

State	Net domestic product			Employment		
	1970-71	1980-81	1990-91	1970-71	1980-81	1990-91
Arunachal Pradesh	59.1	—	—	80.3	64.0	69.6
Manipur	48.0	44.5	44.7	67.0	62.5	64.2
Meghalaya	—	29.1	37.8	78.9	64.1	68.6
Mizoram	—	—	—	—	55.8	67.4
Nagaland	—	17.4	28.7	79.0	73.4	71.9
Sikkim	—	—	—	85.6	64.3	61.4
Tripura	70.0	38.7	45.6	74.3	57.6	61.8
All India	49.2	31.6	36.3	69.7	58.9	60.5

Changes in Land Distribution

The distribution of number of holdings and occupied by various size groups of holding in NEH States show that agrarian structure is still highly skewed (Tables 5.2 and 5.3). The number of holdings less than 2 ha was highest in the country as well as in the region except Nagaland where there is a uniform distribution of holdings. The number as well as area under the holdings less than 2 ha continued to be substantially high in Tripura leading to the lowest average size of holding. Nagaland had the highest average size of holding due to the highest proportion of area and number of holdings above 10 ha. The temporal changes in agrarian structure reveals two distinct patterns: (i) a sharp decline in the area as well as in the number of holdings below 2 ha as in the case of Manipur, and (ii) increase in the area as well as in the number of holdings above 4-10 ha. However, the average size of holding increased from 1.12 ha in 1976-77 to 2.3 ha in 1990-91. This could be because of selling of land by small and marginal farmers as they may seek gainful employment outside agriculture. A decline in area and number of holdings of large farmers in Tripura, Sikkim and Nagaland has been observed implying stronger sub-division of holdings in these States.

Table 5.2: Changes (%) in Agrarian Structure in Green Revolution for NEH States (number of holdings)

State	Year	Number of holdings ('000)				
		0–2	2–4	4–10	Above 10	Total (number)
Arunachal Pradesh	1976-77	—	—	—	—	—
	1990-91	—	—	—	—	—
Manipur	1976-77	84.40	14.11	1.48	Neg.	141.7
	1990-91	83.10	14.79	2.11	Neg.	142.0
Meghalaya	1976-77	64.75	26.56	8.11	0.95	170.2
	1990-91	64.33	26.90	7.60	0.59	171.0
Mizoram	1976-77	—	—	—	—	—
	1990-91	—	—	—	—	—
Nagaland	1976-77	23.72	19.36	32.62	24.49	103.3
	1990-91	23.94	18.31	33.10	23.24	142.0
Sikkim	1976-77	59.49	23.79	13.83	2.89	31.1
	1990-91	69.81	16.98	9.43	1.89	53.0
Tripura	1976-77	83.13	13.34	3.28	0.24	246.6
	1990-91	89.94	8.81	1.26	Neg.	318.0
All India	1976-77	72.64	143.30	10.0[illegible]	2.99	81568.9
	1990-91	78.29	13.06	7.11	1.56	106637.0

Source: Fertiliser Statistics (1977-92), FAI.

Changes in Land-Use and Cropping Pattern

The land-use and cropping pattern in NEH States (Tables 5.4 and 5.5) indicate that the area put to agricultural use was significantly low over the years. It was highest in Tripura (22.9%) and lowest in Arunachal Pradesh (2.04%). It is due to inherent characteristics of terrain, fragility, inaccessibility and marginality of hill and mountain ecosystem coupled with traditional land-use system although the all India level remained more or less stagnant. This shows that land was either made suitable for cultivation through land development activities or land exploitation motive played a part. The area under forests increased from low to moderate and was significantly higher than all India average. However, there was a sharp increase in area

Table 5.3: Changes (%) in Agrarian Structure in Green Revolution for NEH States (area in '000 ha)

State	Year	Number of holdings ('000)				
		0–2	2–4	4–10	Above 10	Average size of holding
Arunachal Pradesh	1976-77	—	—	—	—	4.27
	1990-91	—	—	—	—	3.71
Manipur	1976-77	62.58	30.85	6.26	0.50	1.12
	1990-91	34.77	38.74	23.51	0.33	2.13
Meghalaya	1976-77	33.36	38.73	23.38	4.53	1.74
	1990-91	33.11	38.74	23.51	4.30	1.77
Mizoram	1976-77	—	—	—	—	1.48
	1990-91	—	—	—	—	1.37
Nagaland	1976-77	3.64	7.65	29.06	59.66	7.61
	1990-91	4.03	7.85	30.79	57.44	6.82
Sikkim	1976-77	21.20	26.35	31.62	20.83	2.56
	1990-91	27.03	24.32	27.93	19.82	2.09
Tripura	1976-77	49.26	28.49	14.07	8.18	1.25
	1990-91	62.66	25.00	5.84	6.49	0.97
All India	1976-77	23.52	19.86	30.38	26.25	2.00
	1990-91	32.46	23.19	27.04	17.32	1.56

Source: Fertilizer Statistics (1977-1992), FAI.

under forest in Nagaland. The trend was reverse in case of area not available for cultivation. The cropping pattern in NEH States except Sikkim was predominantly remained foodgrain based. It is interesting to note that in Arunachal Pradesh, Meghalaya, Mizoram, Nagaland and Tripura, pulses and oilseeds gained in area under rice, maize and total foodgrains. It is because of sharp increase in net sown area from 6.80 per cent to 13.65 per cent. Unlike other States, Sikkim experienced drastic decline in area under foodgrain crops owing to diversification towards horticultural crops including spices and floriculture. There was a low to moderate increase in net sown area in most of the NEH States during 1975-76 to 1995-96. The area under forests increased from low to moderate and also significantly higher than all India average (21.98%). The cropping pattern in NEH States except Sikkim where cropping intensity decreased from

Table 5.4: Changes in Land-Use Pattern in NEH States (%)

States	Year	Forest	Area not available for cultivation	Permanent pastures, grazing land and misc. cultivable waste land	Others	Current fallow	Fallow other than current fallow	Net sown area
Arunachal Pradesh	1975	91.33	0.66	—	0.33	0.34	2.09	2.04
	1995	93.79	0.87	—	0.80	0.51	0.66	3.37
Manipur	1975	27.23	65.36	—	1.09	—	—	6.33
	1995	27.23	66.36	—	1.09	—	—	6.33
Meghalaya	1975	36.59	13.74	0.76	6.40	2.45	12.18	7.74
	1995	41.81	10.26	0.76	7.23	2.95	7.36	9.19
Mizoram	1975	61.99	10.04	0.19	0.14	8.71	12.32	3.09
	1995	76.53	10.10	0.19	0.14	8.76	7.76	5.22
Nagaland	1975	17.42	75.74	—	—	—	—	6.80
	1995	56.82	3.69	—	8.41	7.37	6.27	13.65
Sikkim	1975	37.11	39.08	14.43	0.56	—	0.14	8.54
	1995	36.20	38.03	9.72	0.70	0.56	1.27	13.38
Tripura	1975	59.92	5.06	2.86	7.92	0.48	0.20	12.30
	1995	57.77	12.68	—	2.57	0.38	0.95	26.41
All India	1975	21.98	12.97	4.15	1.31	4.09	3.13	46.73
	1995	22.57	13.57	3.63	1.14	4.53	3.28	46.65

Source: Fertilizer Statistics, FAI (various issues).
Basic Statistics of North-Eastern Region, NEC, Shillong (various issues).

Table 5.5: Changes in Cropping Pattern in NEH States (%)

States	Year	Rice	Maize	Total foodgrains	Total pulses	Total oilseeds	Cotton	Cropping intensity
Arunachal Pradesh	1975	52.00	14.85	78.62	Neg.	Neg.	—	113.0
	1995	49.18	14.14	75.12	2.66	9.88	—	131.9
Manipur	1975	84.38	5.19	92.52	2.71	2.00	0.05	145.0
	1995	86.76	2.20	88.74	Neg.	1.70	0.06	176.0
Meghalaya	1975	51.38	7.98	96.46	74.00	3.50	3.69	116.7
	1995	42.59	6.96	53.73	1.30	3.72	3.04	119.9
Mizoram	1975	72.94	5.44	78.97	0.59	—	1.60	106.3
	1995	62.57	7.71	73.76	3.67	6.15	1.01	100.0
Nagaland	1975	58 33	9.21	88.42	3.42	1.84	—	109.9
	1995	63.60	13.16	90.00	7.19	1.25	0.09	108.1
Sikkim	1975	16.59	33.96	72.42	5.38	6.92	—	105.8
	1995	5.60	27.80	53.66	4.12	6.76	—	149.5
Tripura	1975	80.27	Neg.	82.05	1.26	1.86	0.56	155.4
	1995	60.52	0.47	63.62	2.16	2.42	0.35	153.8
All India	1975	22.91	3.50	74.39	14.19	9.82	4.27	121.2
	1995	23.27	3.38	66.50	12.25	13.94	4.85	131.2

Source: Fertilizer Statistics, FAI (various issues).
Basic Statistics of North-Eastern Region, NEC, Shillong (various issues).

106.3 per cent in 1975 to 100 per cent in 1995 remained highly specialised in foodgrains.

Growth in Agricultural Crops Production and Productivity

The growth in production of major crops reflects the performance of agriculture in the region. Compound growth rate (CGR) in the case of rice indicated that there was significant growth in area in Arunachal Pradesh and Nagaland while the other States showed negative growth rates. It is because of diversification of rice crop to other high value crops. Despite decline in area under rice, the production of rice of NEH States except Meghalaya increased significantly due to the improvement in productivity. Mizoram experienced the highest growth rate of 7.5 per cent per annum mostly by increasing productivity. However, the increased production in Arunachal Pradesh was mainly due to area expansion. The area expansion as well as yield increase caused maize output to grow significantly in Arunachal Pradesh, Mizoram, Nagaland and Sikkim. However, in Meghalaya the increase in production was significantly contributed by yield increase. The significant increase in yield of maize in Manipur could not offset the negative growth of area resulting in declined production. In terms of foodgrain production, the significant increase in Manipur, Mizoram, Sikkim and Tripura was due to the improvement in productivity by using high yielding varieties (HYVs). The increase in production of foodgrains in Arunachal Pradesh and Nagaland contributed mostly by area expansion only (Table 5.6).

The production of total pulses in NEH States except Manipur increased significantly due to increase in both area and productivity. The growth in productivity was highest in Sikkim (13.84%) and least in Meghalaya (0.74%). Mizoram increased the area under pulses by 12.41 per cent per annum (Table 5.7) followed by Nagaland (6.16%), Meghalaya (3.58%) and Sikkim (2.90%). On the whole, the foodgrains production in Manipur, Mizoram, Sikkim and Tripura increased significantly mostly due to improvement in productivity.

Both area expansion and improvement in yield resulted significant growth of oilseeds production in Arunachal Pradesh, Mizoram, Nagaland and Tripura (Table 5.7). However, Meghalaya and Sikkim showed low area expansion and yield improvement causing moderate increase in production. Area and yield grew more or less at the same rate in Meghalaya, Mizoram and Sikkim. In Manipur, poor growth

Table 5.6: Trend in Area (A), Production (P) and Productivity (Y) of Rice, Maize and Total Foodgrains in NEH States

(area = '000 ha; production = '000 t; productivity = t/ha)

States	Year	Rice			Maize			Total foodgrains		
		A	P	Y	A	P	Y	A	P	Y
Arunachal	1975	67.6	59.9	0.89	19.3	22.1	1.15	102.2	91.6	0.90
Pradesh	1997	120.0	129.5	1.08	34.5	50.0	1.45	183.3	209.8	1.14
	CGR	2.84	3.25	0.04	3.43	4.54	2.28	3.04	3.82	0.74
Manipur	1975	177.2	276.4	1.56	10.9	23.6	2.17	194.3	303.1	1.56
	1997	157.9	351.7	2.23	4.0	13.0	3.25	161.5	364.8	2.26
	CGR	–0.46	1.64	2.11	–4.97	–2.03	3.1	–0.73	1.45	0.74
Meghalaya	1975	104.3	119.3	1.14	16.2	11.3	0.70	125.4	135.3	1.08
	1997	105.2	150.1	1.43	17.2	24.9	1.45	132.7	186.6	1.41
	CGR	–0.02	–0.16	–0.13	0.27	3.25	2.97	0.10	0.43	0.14
Mizoram	1975	49.6	39.0	0.79	3.7	5.7	1.54	53.7	44.9	0.84
	1997	68.2	110.6	1.62	8.4	15.9	1.89	80.4	128.6	1.60
	CGR	0.01	7.50	5.49	3.15	7.13	3.85	1.26	6.62	5.40
Nagaland	1975	66.5	66.5	1.00	10.5	6.3	0.60	100.8	88.6	0.88
	1997	145.0	187.0	1.29	30.0	30.0	1.00	205.2	239.3	1.17
	CGR	3.30	5.20	1.85	4.80	7.24	2.30	3.65	5.54	1.83
Sikkim	1975	15.1	12.0	0.79	30.9	30.5	0.99	65.9	63.7	0.97
	1997	8.0	15.0	1.76	39.5	55.7	1.41	76.2	103.4	1.36
	CGR	–5.03	0.50	5.67	1.64	3.83	2.20	0.86	3.24	2.37
Tripura	1975	300.2	366.5	1.22	2.00	1.20	0.60	306.9	373.1	1.22
	1997	257.8	535.8	2.08	2.00	1.80	0.90	271.0	547.4	2.02
	CGR	–0.08	1.84	5.67	—	—	—	–0.73	1.93	1.93
All India	1975	39475.0	48470.0	1.23	6031.00	7256.00	1.20	128181.0	121034.00	0.94
	1997	43420.00	82300.00	1.90	6305.00	10852.00	1.72	124068.00	192434.00	1.55
	CGR	0.47	3.02	2.53	0.23	2.51	2.28	–0.17	2.60	2.78

in area could not offset negative growth in yield thereby having a decline in production by 0.32 per cent per annum.

Table 5.7: Trend in Area (A), Production (P) and Productivity (Y) of Total Pulses and Oilseeds in NEH States

(Area = '000 ha; Production = '000 t; Productivity = t/ha)

States	Year	Total pulses			Total oilseeds		
		A	P	Y	A	P	Y
Arunachal Pradesh	1975	5.4	4.2	0.78	0.1	0.1	1.00
	1997	6.5	6.6	1.02	24.1	23.0	0.95
	CGR	—	—	—	6.52	8.63	2.17
Manipur	1975	5.7	2.3	4.0	4.2	1.7	0.40
	1997	4.1	1.6	3.9	3.1	1.5	0.48
	CGR	–5.64	–5.90	–0.29	1.00	–0.32	–1.36
Meghalaya	1975	1.5	1.0	0.67	7.1	3.6	0.51
	1997	3.2	2.5	0.78	9.2	6.0	0.65
	CGR	3.58	2.08	0.74	1.22	1.70	0.40
Mizoram	1975	0.4	0.2	0.50	1.8	0.6	0.33
	1997	4.0	6.7	1.68	6.7	9.1	1.36
	CGR	12.41	25.12	9.86	10.85	21.03	9.15
Nagaland	1975	3.90	2.00	0.51	2.10	0.90	0.43
	1997	16.40	12.60	0.77	28.50	24.80	0.87
	CGR	6.16	9.92	3.56	14.85	17.94	2.70
Sikkim	1975	4.90	3.90	0.80	6.30	4.30	0.68
	1997	6.70	5.90	0.88	9.60	7.60	0.79
	CGR	2.90	3.78	1.34	1.18	1.70	0.50
Tripura	1975	4.70	1.70	0.36	6.90	3.30	0.48
	1997	9.20	5.30	0.58	10.30	8.30	0.81
	CGR	4.48	6.57	1.99	4.34	6.72	2.28
All India	1975	24452	13039	0.53	16922	10607	0.63
	1997	22847	13070	0.57	26013	22016	0.85
	CGR	–0.11	0.95	1.06	2.55	5.03	2.42

Growth in Horticultural Crops and Productivity

The production of horticultural crops has assumed great importance in NEH region in view of diversified topography and conducive agro-climatic conditions. Production of pineapple increased significantly except, a major horticultural crop of the region, increased significantly except Manipur during last seven years. Nagaland experienced the highest growth of 67.84 per cent while Meghalaya registered lowest growth (1.63%). In Arunachal Pradesh and Tripura, the increase was due to area expansion while it was contributed by yield increase in Mizoram and Nagaland. Area expansion and yield increase resulted in increasing citrus production in Arunachal Pradesh and Mizoram whereas in Nagaland and Tripura improvement in productivity increased production. However, there was a significant decline (13.59%) of citrus production in Sikkim despite slight expansion in area. The growth of area under banana and its productivity was positive in all the States except Sikkim indicating expansion of area under banana and intensification of improved technology package in the region. However, the rate of growth varied from State to State. Mizoram recorded highest increased productivity (3.77%) followed by Nagaland. The productivity of potato was also increased in all the states but at a moderate rate while significant expansion of area was observed in Arunachal Pradesh, Manipur, Sikkim and Tripura. The notable feature in Tripura was its remarkability in large-scale production of improved varieties of true potato seeds (TPS). There was expansion of area and productivity of ginger in all NEH States. The highest rate of productivity growth was in Arunachal Pradesh (24.34%) and the least was in Manipur (3.37%).

Growth in Livestock Production

Livestock farming is a way of life of hilly and tribal farming community. It is a profitable venture in NEH States in view of conducive agro-eco-climatic conditions prevailing in the region. Inclusion of livestock component in the hill farming system improved economic conditions of the farmers besides generating employment. Similarly, aquaculture is another profitable enterprise because the region is endowed with large water bodies. So is the case of dairy. The production of milk increased significantly which ranged from 1.49 to 21.6 per cent. The significant increase in milk production in Tripura is mostly attributed to improvement in milk breed. The increase in cattle population varied from 18.9 per cent in Sikkim to 256.99 per cent in Nagaland. There was significant gain in egg production in Nagaland, Manipur, Mizoram and Meghalaya, the growth being more than 6.16 per cent although this increase in egg

Table 5.8: Trend in Milk, Fish, Egg, Cattle, Buffalo and Poultry in NEH States

(*milk and fish '000 t; eggs in lakh; cattle, buffalo and poultry in numbers*)

States	Year	Milk	Cattle	Buffalo	Fish	Eggs	Poultry
Arunachal Pradesh	1975	25.00	168.00	12.00	0.20	170.00	764.00
	1997	44.00	324.00	3.00	2.10	340.00	1187.00
	CGR	1.49	94.86*	–58.33*	10.80	2.70	55.37*
Manipur	1975	51.50	294.00	52.00	1.80	200.00	938.00
	1997	62.00	719.00	114.00	13.70	612.00	3259.00
	CGR	1.45	144.56*	119.23*	10.10	6.58	247.44*
Meghalaya	1975	46.00	477.00	40.00	1.00	220.00	1075.00
	1997	59.00	635.00	33.00	4.10	810.00	1824.00
	CGR	3.12	33.12*	–17.50*	7.60	6.24	169.99*
Mizoram	1975	0.40	49.00	2.00	0.20	30.00	1128.00
	1997	20.00	59.00	7.00	2.70	35.00	1086.00
	CGR	21.60	20.41*	250.00*	9.90	6.71	–3.72*
Nagaland	1975	2.40	93.00	8.00	1.20	106.00	715.00
	1997	46.00	332.00	32.00	2.90	460.00	2164.00
	CGR	19.40	236.99*	300.00*	–4.30	7.83	202.66*
Sikkim	1975	17.00	158.00	5.00	0.10	100.00	221.00
	1997	35.00	188.00	2.00	0.10	170.00	302.00
	CGR	3.88	18.99*	–60.00*	–2.41	2.67	36.65*
Tripura	1975	14.00	592.00	14.00	24.10	19005.80	665.00
	1997	47.00	949.00	20.00	25.80	0.00	2597.00
	CGR	5.83	60.33*	42.86*	10.14	5.15	290.53*
All India	1975	25.10***	180140.00	62029.00	2265.90	8631.50**	160870.00
	1997	71.10***	204516.00	84239.00	5360.00	30150.00**	307071.00
	CGR	4.40	3.53*	35.81*	4.60	6.16	90.88*

* CGR = Compound Growth Rate; ** million; *** lakh tonnes.

Source: Basic Animal Husbandry Statistics, Government of India, New Delhi (various issues).

production could not help in achieving the goal of self-sufficiency and restricting import of eggs from other States of the country. The growth of poultry production was moderate. There was increase in fish production in Arunachal Pradesh, Manipur and Nagaland but this increase could not offset the large-scale supply of fish to NEH region (Table 5.8).

Infrastructure Development and Inputs Used

Infrastructure development and inputs used analysis in relation to output gains points out that the development of irrigation has been quite uneven across the NEH States. An organised effort to harness water resources was not up to the mark in all the States. In Manipur and Tripura where percentage of net irrigated area to cultivated area was high, foodgrains productivity was also more. In general, the growth percentage of irrigated area to net cultivated area remained stagnant due either to increasing net cultivated area or low investment in irrigation development. Fertiliser consumption per hectare was higher in Manipur and Tripura while in other States it was abysmally low although the growth in consumption per annum was quite high. It may be again due to the complementary relationship of fertiliser use with the availability of irrigation facilities. All the NEH States had less fertiliser consumption than all India level of 97.5 kg/ha (Table 5.9).

Table 5.9: Trend in Fertiliser (Nitrogen and Phosphorus) Consumption in NEH States

('000 tonnes)

States	1975-76	1985-86	1996-97	CGR
Arunachal Pradesh	0.08	0.18	0.55	12.25
Manipur	1.34	4.89	13.18	9.90
Meghalaya	1.69	3.07	3.43	3.06
Mizoram	0.10	0.08	0.39	14.75
Nagaland	0.11	0.25	0.85	10.89
Sikkim	1.00	1.17	0.75	9.85
Tripura	0.29	4.22	8.72	15.71
NEH States	4.61	13.86	27.87	10.92
All India	2893.70	8474.10	14308.10	7.36

The percentage area under high yielding varieties varied from 17.84 (Arunachal Pradesh) to 75.19 (Tripura). There was significant growth in rural electrification in NEH States, which varied from 46.3 per cent (Arunachal Pradesh) to 98.0 per cent (Nagaland). However, the number of pump sets per unit area was negligible except Tripura (5.26/'000 ha). Agriculture in NEH State is highly labour intensive as shown by low tractor density per unit area. The existing sloppy and undulating topography restricts the use of tractor in the region. Nagaland, Manipur and Mizoram had comparatively higher growth rate of tractorisation. Only three States showed higher per hectare credit availability than all India average of Rs. 1046. The region has not been able to catch up in farm mechanisation because of its terrain, farming system in vogue, weak extension and promotion efforts. Farming is done by traditional hand tools and bullock drawn wooden implements. Average farm power availability was estimated at 0.9 hp/ha (0.67 kW/ha) with a maximum of 1.44 kW in Meghalaya and a minimum of 0.23 hp/ha (0.17 kW/ha) in Arunachal Pradesh as against 1.15 kW/ha of national average. Green house and covered cultivation has recently shown great potential of increasing production of quality vegetables and flowers. The use of hybrid seeds and planting material developed through biotechnological techniques also help in increasing production and productivity. The use of pesticides was declining in NEH States between 1990-91 to 1995-96, which is a welcome change from environmental and organic farming point of view. The infrastructure development in animal husbandry sector in NEH States has shown wide inter-State disparities. The number of veterinary hospitals/clinics varied from 1 in Arunachal Pradesh to 54 in Manipur (Table 5.10) and other development centres also varied from 1 to 119 (Table 5.11).

Improved technology packages evolved by research institutions of the region had a vital role in enhancing agricultural production. There has been sharp increase in the number of research, teaching and extension institutions in the region in terms of University, colleges and Krishi Vigyan Kendras (KVKs) related to agriculture and allied sectors which are responsible for developing suitable varieties/breed, management practices, manpower training and extension activities. Mention may be made of that the ICAR Research Complex for NEH region including its regional stations and KVKs were the only

Table 5.10: Veterinary Hospital and Dispensaries in NEH States (numbers)

States	Veterinary hospitals			Veterinary dispensaries		
	1985	1991	1998	1985	1991	1998
Arunachal Pradesh	—	—	1	75	90	93
Manipur	49	53	54	86	108	101
Meghalaya	1	4	4	53	55	58
Mizoram	1	3	5	28	39	40
Nagaland	3	4	4	27	27	27
Tripura	5	9	9	28	44	44
All India			7749			15554

Table 5.11: Infrastructure of Agriculture and Allied Sectors in NEH States (as on 2002)

Centre	Arunachal Pradesh	Manipur	Meghalaya	Mizoram	Nagaland	Tripura
University	—	1	—	—	—	—
AC	—	·1	—	—	1	—
VC	—	—	—	1	—	—
CH	1	—	—	—	—	—
FC	—	—	—	—	—	1
RI	2	1	3	1	2	1
ICDP	1	1	2		1	2
AIC	50	154	87	50	7	137
CBBF	27	2	5	7	7	4
PoBF	21	1	10	10	5	5
GBF	—	—	1	1	1	1
PBF	6	5	10	9	8	6
FSPF	—	2	4	3	5	1
SBF	1	—	1	—	1	—
MVD	169	29	79	101	133	371
SWEC	7	—	—	—	—	—
KVK	1	1	2	1	1	2

AC: Agricultural college; VC: Veterinary college; CH: College of horticulture; FC: Fishery college; RI: Research institute; ICDP: Intensive cattle development project; AIC: Artificial insemination centre; CBBF: Cattle buffalo breeding farm; PoBF: Poultry breeding farm; GBF: Goat breeding farm; PBF: Pig breeding farm; FSPF: Fodder seed production farm; SBF: Sheep breeding farm; MVD: Mobile Vetterinary Dispensary; SWEC: Sheep and wool extension centre.

Source: Basic Statistics of NE India, 2000.

centres established in 1975 responsible for R&D activities relating to agriculture and allied sectors for the entire NEH region. At present, the region is having one Central Agricultural University with its campuses on agriculture college in Manipur, veterinary college in Mizoram, horticultural college in Arunachal Pradesh and fishery college in Tripura for human resource development in agriculture and allied sectors, four National Research Centres (NRC) on Mithun, Yak, Pig and Orchid located in Nagaland, Arunachal Pradesh, Assam and Sikkim respectively. Emphasis is now given on watershed management and soil and water conservation measures in different States for augmenting production.

Constraints of Growth

The region is prone to a number of biophysical, institutional and socio-economic problems resulting in subsistence agriculture with low-input-low-risk technology. The major constraints confronting agricultural growth of the region are:

- The acidic soils of the region result in aluminium toxicity and associated nutrient problems causing poor plant growth. The high rainfall and humidity not only create favourable environment for prolific occurrence of diseases, pests and weeds but also create problems to control them. The high rainfall and cloudy sky reduce the total sunshine hours needed for photosynthetic activity. The humid climate also favours high incidence of animal health resulting in poor health and low productivity.
- The continuous short cycle of shifting cultivation, the dominant farming system of the region, in some areas has left the hills barren without soil and badly degraded fields are infested with imperata (*Imperata cylindrica*) grass. The degradation of shifting cultivation has resulted in declining upland agro-ecosystems.
- The ownership of land by community or the village chief and the prevailing land tenure system often act as disincentive for development and maintenance of land for cultivation.
- The cultivation along the hill slope without proper conservation measures and rapid loss of tree cover from

mountains has rendered them susceptible to accelerated soil erosion, landslides, loss of habitat and genetic diversity.

- High level of diversity in the upland farming systems in different agro-climatic zones does not allow the uniform package of practices for agricultural development.
- Limited opportunities for expanding arable farming to maintain fragile hill ecosystem.
- Limited availability or access to improved varieties of seeds/ planting materials/breeds.
- High dependency of hill farmers on rice production and low diversification of cropping systems.
- Lack of integration and coordination between different line departments in the promotion of integrated development strategies.
- Inadequate extension and training support services with little client oriented or participatory extension activities resulting in communication gaps between extension agencies and farmers.
- Lack of adequate service facilities like timely credit, input supply and marketing to support agricultural development activities.
- Dependent attitude of farmers who have continuously benefited from Government sponsored subsidised schemes.
- Relative isolation and poor accessibility of most of the village settlements.
- Transport and communication bottleneck and the resultant high transport cost in the region limit scope for exploitation of development potential in areas like plantation and horticultural crops.
- Use of improved farm tools and machinery are not in vogue.

Critical Gaps

- Alternative farming systems to replace *jhuming* and high yielding varieties developed by ICAR Research Complex have to be taken up as a priority in a stakeholder and participatory mode in order to bridge the gap between total food production and availability.
- Much more emphasis on the importance of horticulture

including floriculture as well as post-harvest technology has to be laid from export potential point of view.

- Replication of tested watershed based systems in the farmers' fields and encouragement of community participation in watershed based management needs to be taken up.
- The region suffers from adequate processing and marketing bottlenecks of agri-horticultural products and, therefore, marketing and financing agencies need to be developed.
- Conservation of the rare crop and animal genetic resources under *in situ* and *ex situ* condition has remained a gap.
- The correlation between bamboo flowering and large-scale multiplication of rodents leading to famine condition in Mizoram is yet to critically assessed.
- A large number of lesser-known crops and fruits of economic importance available in the region have not yet been systematically surveyed and documented.
- Production and availability of farm tools and implements, which help in human drudgery in the remote areas, are yet to be achieved.
- Assessment of different farming/land-use systems to improve the crop husbandry practices through integrated plant nutrient supply system is another critical area.
- Easy availability of improved pig, poultry birds, animal feeds and adequate facility for diagnosis and control of livestock diseases have to be ensured for the development of animal husbandry sector.
- Social science research in terms of agricultural development in the region, by and large, has remained a critical gap to impact analysis of changing scenario of agriculture.

Future Potential

The following items have good export potential provided international quality control is maintained in both production and processing with value addition:

- Ginger and turmeric
- Glutinous rice
- Cut flower and orchids
- Bamboo shoots and other diversified bambooo products

- Ornamental fish
- Processed pineapple and jackfruit products.

Adoption of improved technology was limited in early seventies which affected yield rate. With the development of transfer of technology (ToT) in extension approach, the production and productivity of agricultural crops have been increased. The training and demonstration programmes conducted by various institutions have undoubtedly created a zeal for scientific cultivation among the practicing farmers towards increased crop production, maximum utilisation of land, orchard raising and farm forestry, livestock rearing, dairy farming and other diversified farming. These measures facilitate an equitable growth and development in the region.

6

CHANGING AGRICULTURAL SCENARIO IN NORTH-EAST INDIA

— P.C. DUTTA and B.C. PRADHAN

> In spite of the population explosion, it is possible to substantially augment the carrying capacity of the land, double or even triple the foodgrain production and achieve rapid industrialisation to provide economic security and improve the quality of life of our people, provided we appeal to the development in Science and Technology and pursue the strategy of sustainable integrated development.[1]

The North-Eastern Region or its popular shortened form NER comprises seven States of Indian territory forming 7.8 per cent of the total land area and about 4 per cent of the total population of the country. More than 70 per cent of total geographical area of NER is covered by hills and about 3 million hectares are estimated to be under soil erosion hazard as a result of practice of *jhum* cultivation. In Assam alone, 83.2 per cent of area suffered from erosion of slight (35.3%), moderate (37.7%), severe (10.0%) and very severe (0.3%) intensity. Traditionally, agriculture has been the mainstay of people of this region but the agro-based economy fails to flourish as it should have due to lack of proper involvement and utilisation of technological aids. As a result, the stamp of 'backwardness' has been attached to this region suffering from food scarcity while the country moves ahead from its target production to food surplus in different phases of the post-Independence era. Increasing the yield of crop in a complex system and in a environmentally positive manner is a challenge that will not be easy to meet in a place like NER.

Since the planning period started, stress had always been laid on the production of foodgrains to meet the requirements of the increased population. In this respect NER cannot be compared with all other States of India as growth of population is most alarming in this region due to immigration from neighbouring border countries and the flow of population to this region from other States of Indian Union. This problem of food insecurity can be linked with the backwardness in agricultural development.

Our study began as an attempt to evaluate the on going debate regarding the food prospects of India (Bhalla and Hazell, 1997; Kumar, 1998; Bansil, 1999; Bhalla, Hazell and Kerr, 1999). This debate is usually depicted as hinging around the issue of future food demand elasticity and the extent to which rise in per capita income will be translated into increased demand for livestock products. As NER is a food deficit area, it cannot be compared with the other States of India, for its peculiar socio-economic and demographic features.

Development has been appropriately conceptualised as a process, which improves the quality of life (Narain *et al.*, 2003). The programmes of development have been taken up in the country in a planned way through various Five Year Plans. The main objective of these programmes is to enhance the quality of life of people as well as effecting improvement in their social and economic well-being. The economic growth and uniform regional development are the basic objectives of developmental programmes. The Green Revolution in agricultural sector and commendable progress in the industrial front have increased the overall total production in the country, but there is no indication that these achievements have been able to reduce substantially the regional inequality in the level of development. Although resource transfers are being excuted in the backward region of the country, it has been observed that the regional disparities in terms of economic development is not declining over time.

Dutta (2003) observed that the States of NER are suffering from the paucity of funds to improve their socio-economic backwardness. The States of NER are far behind from other States of the Indian Union so far as the basic amenities of life are concerned. He analysed and concluded that the economic development of a country may not be possible without regional development and equitable distribution of wealth and income of the nation among the different parts of the country. Inequitous distribution of national wealth and resources of the country causes regional disparities by giving rise to various types of socio-economic problems.

T. Dyson and A. Hanchate (2000) concluded that population and food are closely related subjects; yet they are often tackled in relative isolation from each other. Levels of infectious diseases can have a major and independent influence upon a population's nutritional status.

M. Swaminathan (1999) observed that in some States, the private traders had lower prices for foodgrains than the agencies of PDS (Public Distribution System) because of the transportation. The author suggested that an expansion of the geographical coverage of procurement and promotion of decentalised procurement may help to contain the rise in procurement prices.

R.K. Samanta (1994) highlighted the various issues and perspective of agriculture and food production in NER with reference to the tribal culture and eco-friendly activities by which sustainable development can be achieved.

Misra and Rao (2003) examined whether trade liberalisation helps to accelerate growth in agriculture and reduce poverty. He observed that the adverse impact of trade policy operating through terms of trade on the rural poor may not entirely be attributed to the process of globalisation of Indian economy. It may be noted that factors such as capital stock, net area sown and private investment play very important role in the positive aggregate crop output response. He analysed and concluded that the slower increase in agricultural production during the nineties may be attributed to declining real public investment in agriculture and not the failure of trade and agricultural price policies.

Dutta and Pradhan (2000) observed that the most important problem which creates the hindrance on the way of agricultural development is the problem of population explosion in the region. Although India has emerged as a food surplus State in the world, the NER is still suffering from food deficit. In almost all the States, there are substantial urban-rural and inter-district variations, along with the marked variation between tribal and non-tribal population within the district, for which it is always a difficult task to measure the nutritional standard of the population living in a particular region and the case of NER is highly apparent.

In fact, despite several measures taken by the government, the purchasing capacity of the people of North-East has not increased up to the level our planners and policy-makers desired.

It was pointed that the poorest section of the society has been badly affected in the States of NER. The extent of degree of inequality in the level of living in the States of NER is much more than what is taken to guide some of the decision-making elements in connection with resource transfer from central to these North-Eastern States (Dutta and Choudhury, 1995, 1996).

Rice is the most important crop grown in India constituting about 44 per cent of the total foodgrains accounting for about 24 per cent of the total cropped area.

The change in external environment for agriculture came with the setting up of the World Trade Organisation (WTO) on January 1, 1995 and with the inclusion of agriculture under the multilateral trade discipline for the first time. With this development and signing of Agreement of Agriculture, it became inevitable that India has no option but to globalise its economy including agriculture (Kalirajan *et al.*, 2001).

After the introduction of economic reforms in India, the government neglected the importance of agriculture to the Indian economy and agricultural production started decelerating. We must realise that in India agriculture is a key to rapid and sustained economic growth. The Finance Minister seems to have realised this in his recent budget speech where he indicated that there is a need for agricultural revolution in India, but the strategy for the revolution is not planned out as yet (Kapila, 2002).

Agriculture plays a crucial role in the development of Bangladesh national economy, accounting for 32 per cent of its gross domestic product, contributing about 24 per cent of its exports and employing over 60 per cent of its labour force. The rate of growth of agriculture in Bangladesh was generally low at about 2 per cent per annum between 1975 and 1994, compared to overall economic growth of 3.6 per cent (Faruqee, R., 1998). The author noted that the future of agriculture will depend on how the outstanding issues of key instruments such as agrarian structure, land reform and environment, public food distribution system, government policy, etc. are addressed.

He suggested that the future growth of agriculture would require policy changes in four key areas. The important need is emphasised to target the food insecurity and to remove untargeted subsidies that are appropriated, at least partly, by the rich.

The Family Welfare Programme is also an indirect nutrition programme since by reducing family size, it will not only increase the per capita availability of food but also result in all round facilities including health care. The women play a vital role to maintain a perfect balance between human activity and environment. Their participation in both rural and urban development will help to maintain a sustainable eco-system as the women are continuously and closely in contact with the forest for fuel and other activities. Thus to check the population growth and to save the forest from degradation, the employment of the women by proper training is necessary (Dutta, 2002).

In view of the above context, the present study is an attempt to focus how the problems of agricultural development can be linked with the social system and the governmental policies of the region. The primary objective of agricultural development is to satisfy the essential materials needs of the society, but the same cannot be achieved at the cost of local environment and the culture of the people. Hence, care has been taken to study the tribal way of life and its impact on the agricultural development of the region.

To analyse the agricultural backwardness of the region, some of the social indicators have been chosen. As it is very much customary in this region to blame the rise of insurgency as the major factor which impedes the development of the region in general and the agricultural development, in particular, the study rejects the notion as some of the States of the region made rapid progress in agriculture during the period of insurgency and rise of terrorism.

Finally, the study, based on statistical data and other secondary sources, is concluded with some remedial measures and suggestions for the future planners of the country as well as people of the North-East to cope with the currents of new millennium so that a hunger free North-East can be achieved.

Agricultural Development and Allied Problems

Indian economy is pre-eminently agrarian and North-East India is no exception to it. Agriculture occupies a key position in the economy of NER because of its contribution to:

(a) overall economic growth through supplies of food, raw materials and export,
(b) source of livelihood for over 70 per cent of population, and
(c) providing a large market for non-agricultural goods and services.

Agriculture in NER has considerable social and political influence. Agriculture has a vital role to play in the economic development of NER, which is a key factor in the economic growth in absence of mushrooming growth of industries as seen in other parts of India especially in western and southern States. NER produces a wide array of crops although the food crops preponderate over other crops accounting for maximum total cultivated area of the entire region. Among food crops, rice is the major crop followed by maize and pulses. But the slow growth in agricultural sector results in the slow growth of North-East economy. Again, the policy-makers are more keen on accelerating industrialisation than the development of agriculture.

However, the agricultural backwardness of the region is clearly visible in the context of present day India. While other States of India like Punjab, Haryana, West Bengal are marked by food surplus in the production of rice, wheat, fish, etc., the people of NER are importing two million tonnes of foodgrains from other regions of the country at a higher prices.

It is very customary in this region to blame the rise of terrorism as one of the major factors which impedes development of the region in general and the agricultural development in particular. However, during the period of insurgency, the studies show that States of Nagaland, Manipur, Mizoram made rapid progress in the production of foodgrains.[2] Hence, we cannot generalise the problem of terrorism as the greatest hindrance in the development of agriculture and in the production of foodgrains of the region for which the security of life is always threatened.

Food originates from efficient, effective and environmentally benign production technique that conserve and enhance the natural resource base of crops and animal husbandry, forestry and fishery. In Assam rice production this year is surplus to the tune of 41 thousand tonnes. The problem lies in the fact that there is high level of poverty and lack of purchasing power which reduce the ability to access to food and adequate nutrition. However, economic and ecological access will remain as the main challenge at present.

The impending food crisis over the world is due to increasing population, increasing purchasing power of affluents leading to the consumption of more animal products, increasing damage to the ecological foundation of agriculture, depleting per capita availability of land and water and the absence of technologies that can further enhance the yield of potential of major food crops.

The most important problem which creates hindrance on the way of agricultural development is the problem of population explosion in the region. The total population of the region is about 38.5 million representing about 3.75 per cent of India's total population in 2001. The region's growth rate of population is much higher than the national average. It is striking to note that India's total population increased by 51.80 per cent in the period 1951-71, 54.39 per cent in 1971-91 and 21.35 per cent in the period 1991-2001 whereas for the NER the respective growth rates are 90.86, 62.11 and 22.02 per cent. At the same time the share of population of NER, in comparison to India, increased from 2.84 per cent in 1951 to 3.75 per cent in 2001 (Table 6.1).

Table 6.1: Population Trend in North-East Region (in '000)

State	1951	1961	1971	1981	1991	2001
Arunachal Pradesh	—	337	468	632	865	1091
Assam	8029	10837	14625	18041	22414	26638
Manipur	578	780	1073	1421	1837	2389
Meghalaya	606	769	1012	1336	1775	2306
Mizoram	196	266	332	494	690	891
Nagaland	213	369	516	775	1210	1989
Tripura	639	1142	1556	2053	2757	3191
NER	10260	14501	19582	24752	31547	38495
India	361088	439235	548160	683329	846303	1027015
Decadal growth Rate of NER	—	41.34	35.04	26.40	27.45	22.02
Per cent of NER with respect to India	2.84	3.30	3.57	3.62	3.70	3.75

The population explosion caused by immigration poses serious threat on the natural resources of the region and the availability of cultivable lands for the people of NER. As a result, agricultural production could not cope with the growing population of the region for whom the foodgrain production is short by 2 million tonnes of demand in this region.

The absolute number of the migrants was more than 91 thousand in 1901. Since then it has been steadily increasing and reached at nearly 745 thousand in the census year 1991. The share of migrants from Bangladesh was more than 87 per cent in pre-Independence era. In post-Independence era it reached at 94 per cent in 1951—the highest in the periods 1901-1991. From 1961 onwards the share started to decline and became around 89 per cent in the census year of 1991. It is also clear from the statistical information that Bangladesh was followed by Nepal in sending migrants to the NER of India. Again, the number of migrants who came from other States of India varied between 0.5 million to around 0.8 million during the pre-Independence era. But it declined to around 0.4 million in 1951 and reaching slightly more than 0.5 million in 1991.[3] In recent years the flow of immigration again shows an increasing trend.

Since North-Eastern society is dominated by the tribal population, the reference may be made to the tribal method of cultivation. 19.91 lakh hectares (83.73%) of land in the North-East is under shifting cultivation or *jhuming* a practice once considered as a farmer friendly has now become an ecological menace. Rapid deforestation over the years 1987-1997 due to the shifting cultivation in the region have reduced the total forest cover amounting to 1312 sq. km. (Table 6.2). An inevitable consequence of deforestation is increased run off rain water precipitation and extensive soil erosion resulting in unacceptable deterioration of top soil, degradation of land and sedimentation of water bases resulting in desertification and frequent flooding. The high rate of soil erosion in deforested areas in our country ranges from 10 tonnes per hectare in the plains to about 30 tonnes per hectare in the North-Eastern hilly region,[4] 10 to 50 times larger than in the forest areas.

The Union Government has sanctioned Rs. 68 crore to the eight North-Eastern States during the past three years for modernising the forest protection force, forest protection system and strengthening the forest management. Tripura and Assam got the maximum amount of Rs. 16 crore each while the other States of the region received 5-9 crore for this purpose. Among the eight NE States, including Sikkim, the forest cover had increased in Arunachal Pradesh and Tripura (*NE News Letter*, Vol. 5, No. 4, April, 2003).

Table 6.2: Forest Cover in NER of India (sq. km.) During the Period 1987-1997

State	1987	1989	1991	1993	1995	1997	Total change over decade	Total geo-graphical area of land (in sq.km.)
Arunachal Pradesh	60500	68783	68757	68661	68621	68621	–162	83578
Assam	26386	26058	24751	24508	24061	23824	–2562	78523
Manipur	17679	17885	17685	17621	17558	17318	-567	22356
Meghalaya	16511	15690	15875	15769	15714	15657	–854	22489
Mizoram	19092	18178	18858	18697	18576	18775	–317	21087
Nagaland	14351	14356	14321	14348	14291	14221	–135	16527
Tripura	5743	5325	5535	5538	5538	5346	-397	10477
NER	160262	166255	165777	165142	164359	164943	–1312	255039
India								3280483

The environmental problems caused by increased sedimentation has extensively affected the capacity of the reservoirs as well as the aquatic eco-system in the inland waters just as soil erosion, in addition to loss of fertility, have drastically reduced the recharging capacity, resulting in severe depletion of underground water. Extensive deforestation has resulted in increased carbon dioxide concentration in the atmosphere. The changes in climate and rain pattern are gradually setting in because of deforestation. Water and soil erosion have affected the maximum lands used for agricultural purpose.

The climatic fluctuation and changes in the rain pattern have further compounded the problem of providing adequate food security to our increased population. The gross neglect of environment also resulted in polluting not only river water but also surface water, making them fertile breeding grounds of many communicable diseases due to indiscriminate discharge sewage, dumping of solid wastes, etc. Further, the problem arises due to common ownership of the property right which adversely affects not only the loss of fuel and food but also fodder to a larger extent.

In the entire North-East, the valley districts of Manipur are among the most productive zones in the production of paddy. However, the same cannot be said to be true about the districts of Assam. Moreover, the average yield of paddy in Assam is also lower than the regional yield. Lack of assured irrigation seems to be the major factor behind it. The productivity of paddy in Manipur-Tripura Valley and Mizoram is higher than the regional average and also above Assam which alone accounts for more than 70 per cent of area and output in the region. Mizoram has achieved remarkable growth in the yield rate of paddy since around the middle of 1980. It is due to increase in fertiliser consumption and the new land-use policy in Mizoram. Also it is observed from Table 6.3 that while the Government of Assam fixed the amount for expenditure in agricultural

Table 6.3: Agricultural Development in NER

(*Rupees in lakhs*)

State	Unutilised amount of money	Amount fixed for expenditure as on 1.4.98	Amount disbursed	Expenditure
Arunachal Pradesh	421.96	980.32	327.05	94.11
Assam	1440.65	1557.82	365.65	58.67
Manipur	919.52	1538.84	474.90	223.64
Meghalaya	334.20	878.91	523.06	74.01
Mizoram	204.72	1434.17	941.18	283.48
Nagaland	415.22	1799.22	1333.18	574.92
Tripura	537.38	1015.77	438.90	308.34

sector as Rs. 1557.82 lakhs as on 1.4.1998 but disbursed only 23.60 per cent while only 15.95 per cent of the disbursed amount got actually utilised. The same is true in the States of Arunachal Pradesh and Meghalaya. On the other hand, the Government of Tripura and Manipur have not only disbursed more but also actually utilised 70.25 per cent and 47.09 per cent respectively on agricultural sector. Perhaps corruption at the governmental level and red tapism of the bureaucrats are responsible for this problem in the States of NER. Although the paddy productivity is the lowest in the State of

Arunachal Pradesh, the yield of oil seeds and pulses is the highest in the State along with Meghalaya, Mizoram and the hill districts of Manipur. As the region is classified into hill areas and plain areas, the cropping pattern in hill districts is more varied and not similar to the plain areas. The development of infrastructure facilities like roads, banking, health and communications are perhaps responsible for the above gaps.

The low productivity of foodgrains explains the shortage of food in the NER and is responsible for poverty of agricultural masses. Agricultural productivity is in inverse proportion in NER to the number of people engaged in it. There is excessive pressure on land resulting in small uneconomic and fragmented holdings. Such small and scattered holdings coupled with defective land tenure system are bound to keep the agricultural productivity very low.[5]

Animal husbandry and Dairy production is an important subsidiary activity of the people in the North-East. Meat consumption is very high but region draws upon supplies from outside to meet its growing demand. As against the all-India average of 29 cattle for 100 persons, NER has 36 cattle for 100 people. The average number of pigs per 100 people in the North-East is 53 as against the all-India average is 1 pig. The same is true of poultry farming.

However, the production of milk and eggs is less compared to population of cows and buffaloes and poultry birds and ducks. The traditional food habit of the people of the region and the poor management practices of the government are perhaps responsible for this.

Fish is an important item of diet of the people of North-East. According to an estimate, 2,03,617 hectares of the cultivable water area is available for fish farming in the North-East. This is apart from 1,61,713 hectares of water spread area in lakes and beels. However, scientific pisciculture has not yet developed in the region and the availability of fish is much less in comparison to its demand.

A sizeable area in the region is in plantation of tea. Its share in the total production of tea in the country in the year 1980 was 53.44 per cent. Many parts of the region are showing good potential for the cultivation of rubber, coffee and a variety of horticultural crops. The region has a total cultivable area of 6.13 million hectares out of which net area sown is only 0.80 million hectares. However, the above production have not reached the target due to various socio-economic problems of the region.

Nutrition and Food Security

It is now a well-known fact that nutrition plays an important role in physical efficiency of the people and act as a device to determine the food security in a country and in different parts of the country. India is a vast and varied country. There are large differences in per capita income, availability of the foodstuffs, purchasing power, dietary habits, habitual consumption of foodstuffs, health and nutritional status between different States in the country and even in different areas of a State. In almost all the States, there are substantial urban-rural and inter-district variations along with the marked variation between tribal and non-tribal population within the district, for which it is always a difficult task to measure the nutritional standard of the population living in a particular region and the case of NER is highly apparent. It is always a fashion in India to relate food problem with income or private household consumption expenditure. The nutritional value of food items is implicitly or explicitly assumed to be constant. The income elasticity of poor households is supposed to be always less than unity. The income elasticity is sometimes calculated for the sake of deriving relationship with food expenditures. The nutrient intake, household income and working capability of individuals in the household are interrelated. The expenditure incurred on food items is an indication of nutrition intake. Each food item has some calorie value and, therefore, total calories arise from those purchased and consumed as indicated through private consumption expenditure of a household. At the same time it should also be possible to say that lower the consumption expenditure of a household, lower is the calorie intake and that is why intake of calories determines the food poverty. The dietary standards differ from one country to another. The recommended intakes of nutrients of a nation differs from those of international norms set up by FAO and WHO. The term used to describe dietary standard is "Recommended Dietary Allowance (RDA)".[6] Table 6.4 shows the quality of items (Recommended Dietary Requirements) should be consumed by the persons of different sex and age groups.

Table 6.4: Recommended Dietary Requirements in Grams

Food items	Adult male			Adult female			Children		Boys	Girls
	(1)	(2)	(3)	(1)	(2)	(3)	(4)	(5)	(6)	(6)
Cereals	460	520	670	410	440	575	175	270	420	380
Pulses	40	50	60	40	45	50	35	35	45	45
Leaf Vegetables	40	40	40	100	100	50	40	50	50	50
Other Vegetables	60	70	80	40	40	100	20	30	50	50
Roots & Tubers	50	60	80	50	50	60	10	20	30	30
Milk	150	200	250	100	150	200	300	250	250	250
Oil & Fat	40	45	60	20	25	40	15	25	40	35
Sugar & Jaggery	30	35	55	20	20	40	30	40	45	45

(1) Secondary Worker (2) Moderate Worker (3) Heavy Worker
(4) 1-3 years (5) 4-5 years (6) 10-12 Years.

In the North-Eastern hill areas food production is subject to wide fluctuations due to shifting cultivation. Hence, estimate of production may not give the true picture. However, during the last two decades per capita production has been declining in the States of Assam, Manipur, Meghalaya and Mizoram and are lower than the national average except in the State of Arunachal Pradesh. While the population is rapidly increasing in the States of the region, it is observed that in case of Assam, the per capita food production decreased from 158 kg. in 1972-73 to 154 kg. in 1990-91 which further decreased to 139 kg. in 1997-98. The respective figures for Meghalaya are 120 kg., 87 kg. and 83 kg. only (Table 6.5).

It is interesting to note that on a global basis, only 8 per cent increase in foodgrain production in the last 50 years, has been through expansion of cultivable area, the rest 92 per cent coming from improved irrigation techniques and use of high yielding varieties. About 46 per cent of the global foodgrain production comes from irrigated area, which has increased from 80 million hectares in 1950 to 240 million hectares in 1993. It will not be possible to increase the total annual foodgrain productivity in the country beyond 250 million tonnes, with the present agricultural practices, which totally falls short of the requirement to produce a minimum of 400 million tonnes by the year 2050 to provide adequate food security to the projected 1.8 billion population.[7]

Table 6.5: Per Capita Foodgrain Production, Consumption of Electricity and Percentage of Households having Electricity, Safe Drinking Water in N.E. States

	1*	2*	Per capita foodgrain production (in kg.)				3*
	1996-97	1991	1972-73	1980-81	1990-91	1997-98	
Arunachal Pradesh	80.8	40.85	104	207	253	186	70.02
Assam	107.6	18.74	158	150	154	139	45.85
Manipur	127.9	50.92	159	205	155	152	38.72
Meghalaya	134.5	29.16	120	116	87	83	36.16
Mizoram	127.8	59.20	244	70	118	123	16.21
Nagaland	88.0	53.42	107	136	160	150	53.37
Tripura	80.4	36.93	116	196	187	152	37.18
India	338.5	42.37	172	190	208	198	62.30

1* per capita consumption of electricity (in kWH).
2* percentage of households having electricity.
3* percentage of households having safe drinking water.

A look at Table 6.6, makes it clear that the performance of all the States of NER excluding Arunachal Pradesh and Tripura, are below the national average (522.7 gms. per capita per day). In case of per capita per day availability of pulses, it is observed that all the states of NER are below the all-India average over the period 1983-1999.

Table 6.6: Per Capita Availability of Cereals and Pulses (in gms.) per Day in NER

State	Cereals				Pulses			
Year	1983	1988	1994	1999	1983	1988	1994	1999
Arunachal Pradesh	569.8	675.4	653.9	446.6	—	—	15.3	16.8
Assam	365.8	359.5	410.2	373.9	7.10	7.71	6.72	7.69
Manipur	418.5	461.0	492.4	476.4	2.93	—	—	—
Meghalaya	293.8	235.1	208.2	225.9	3.88	4.03	3.54	3.07
Mizoram	179.1	233.2	398.6	395.8	0.52	4.39	35.2	28.6
Nagaland	392.6	247.2	432.1	423.6	3.88	9.58	19.82	22.1
Tripura	535.6	475.8	457.1	391.9	3.02	4.13	5.91	3.9
India	448.4	443.6	520.5	522.7	45.2	37.6	40.5	41.1

Table 6.7 gives a picture of per capita per day availability of foodgrains production where the same picture is prevalent. It is observed that Arunachal Pradesh, the only State in NER tried to maintain its superiority till 1994 but declined thereafter.

Table 6.7: Per Capita per Day Availability of Foodgrains, Growth Rate of Population and Production of Foodgrains

State	Per capita per day availability of foodgrains (in gms.)				Annual average growth rate of population and production of foodgrains			
	1983	1988	1994	1999	1980-90		1991-2000	
					Population	Production	Population	Production
Arunachal Pradesh	569.75	675.42	669.18	463.33	3.1	4.2	3.1	−1.8
Assam	372.88	367.25	416.92	381.58	1.5	1.3	1.2	1.5
Manipur	421.44	461.00	492.35	476.35	2.6	−1.7	2.5	2.2
Meghalaya	297.63	239.21	211.78	228.99	2.8	−0.4	2.8	2.0
Mizoram	179.57	237.55	433.68	424.38	3.4	8.3	3.3	7.4
Nagaland	396.52	256.81	451.87	445.60	4.3	3.7	4.4	4.5
Tripura	538.56	479.93	463.04	395.25	3.4	1.2	2.9	0.2
India	493.60	481.14	560.98	563.84	2.1	1.8	1.9	1.8

Thus food insecurity is a matter of great concern not only to the people of NER but also to the policy makers and planners of NER. It is further observed that no proper care has been taken for the development of agriculture to increase production by which the living condition of the people of NER may be a better one. Modern technologies, irrigation facilities, scientific and modern equipments are not being properly used for the real development of agriculture even in the age of globalisation and liberalisation.

Conclusion and Suggestions

The continuous neglect of soil health and fertility exhaustive methods of cultivation must be rectified through appropriate soil conservation methods and efficient use of organic and inorganic plant nutrients through improved soil management practices. Though reduced use of fertilisers is advocated for environmental effects, the use of inorganic fertilisers is a must in the present moment where soil fertility is low and loss of soil and nutrients are more and large section of population are poor and insecure in food supply.

Distribution of improved varieties of crops, arrangement for fertiliser and proper irrigation facilities, availability of these inputs, demonstration and training for effective and efficient use of organic and inorganic fertilisers in various farming systems through Integrated Nutrient Management System and plant protection through Integrated Pest Management by bio-pesticides, etc. can be effectively managed in these areas. Besides long-term property rights to land, access to credit, marketing facilities of produce and investment in roads and rural transportation system are to be efficiently managed.

Educated youths of the villages must be attracted in farming through spreading science-based precision farming techniques which are both intellectually stimulating and economically rewarding. Increased investment in effective. Agricultural research through private sector, NGOs and other agencies is most important in NER in particular to achieve household food security and nutritional security.

In the Indian Council of Agricultural Research (ICAR) vision 2020 strategic plan to achieve food and nutrition security it is proposed to address the issues of bio diversity, natural resource management, farming system approach, rain-fed agriculture, post-harvest management, farm mechanisation, information networking and effective partnership with other public and private institutions.

It is necessary to improve all the factors employed in agricultural *viz.*, land, labour, capital and enterprise. Poor quality of seeds, non-adoption of modern technologies, non-use of plant protection measures, out-of-date implements, lack of adequate or biased finance (agricultural credit), absence of productive investment in land due to defective land tenure system, poor communication between research and extension or development agencies and lack of irrigation facilities are undoubtedly promoting low agricultural productivity in the region.

NOTES

1. U.R. Rao (1996). General Presidential Address in 83rd Indian Science Congress, Patiala.

2. B.B. Dutta (1994). "Insurgency and Economic Development in India's North East" in R.K. Samanta, (ed.), *India's North East—The Process of Change and Development*, B.R. Publishing Corporation, New Delhi, pp. 1-38.
3. S. Bandyopadhyay and D. Chakrabarty (1999). "Migration in the North Eastern Region of India During 1901-1999: Size, Trend, Reasons and Impact", *Demography India*, Vol. 28(1), pp. 75-97.
4. U.R. Rao (1994). "Space for Sustainable Development", G.B. Panth Memorial Lecture, Guwahati.
5. B.K. Sarma and P.P. Pal (2000). *Food Production, Food Problem and Public Distribution System in North East India*, Paper presented in the National Seminar on Food, Nutrition, Food Security, Public Distribution System, NEICSSR, Shillong, 27-28 November.
6. R. Gedam (1998). *Poverty in India*, Deep & Deep Publications, New Delhi, pp. 386-420.
7. V.R. Rao (1996). General Presidential Address in 83rd Indian Science Congress, Patiala.

REFERENCES

Bansil, P.C. (1999). *Demand for Foodgrains by 2020 A.D.*, Observer Research Foundation, New Delhi.

Bhalla, G.S. and P. Hazell (1997). "Foodgrains Demand in India to 2020", *Economic and Political Weekly*, December 27.

Bhalla, G.S., P. Hazell and J. Kerr (1999). *Prospects for India's Cereal Supply and Demand to 2020, Food, Agriculture and the Environment*, Discussion Paper 29, International Food Policy Research Institute.

Dutta, P.C. (2002). "Development Social Forestry in Assam of North Eastern Region", *IJRS*, Vol. 34, No. 2, pp. 52-62.

Dutta, P.C. (2003). "Regional Disparity and Its Consequences in North Eastern Region of India", in A.C. Mahapatra and C.R. Pathak (Eds.), *Economic Liberalization and Regional Disparities in India—Special Focus on the North Eastern Region*, Star Publishing House, Shillong, pp. 179-82.

Dutta, P.C. and H. Choudhury (1995). "Per Capita Household Consumption Inequalities in North Eastern Region (NER)", *Indian Journal of Regional Science*, No. 27(1 & 2).

Dutta, P.C. and H. Choudhury (1996). "A Note on Inter State Variation in Per Capita Non-food Household Items of Consumption in the North East India", *Indian Journal of Regional Science*, No. 28(2), pp. 101-110.

Dutta, P.C. and B.C. Pradhan (2000). *Agricultural Development and Security of Life in North East India*, Paper presented in the National Seminar on Food Security, Nutrition and Public Distribution System, NEICSSR, Shillong, Nov. 27-28.

Dyson, T. and A. Hanchate (2000). "India's Demographic and Food Prospects—State Level Analysis", *Economic and Political Weekly*, November 11.

Faruqee, R. (1998). *Bangladesh Agriculture in the 21st Century* (Ed.), The University Press Ltd. Dhaka, Bangladesh.

Kalirajan, K.P., G. Mythili and U. Sankar (2001). *Accelerating Growth through Globalization of Indian Agriculture* (eds.), Macmillan India Ltd., New Delhi.

Kaplia, Uma (2002). *Indian Economy Since Independence* (Ed.), Academic Foundation, New Delhi.

Kumar, P. (1998). *Food Demand and Supply Projections for India*, Agricultural Economics Policy Paper 1998-2001, Indian Agricultural Research Institute, New Delhi.

Narain, Prem, S.D. Sharma, S.C. Rai and V.K. Bhatia (2003). "Evaluation of Economic Development at Micro Level in Karnataka", *Journal of Indian Society of Agricultural Statistics*, 56(1), 52-63.

Misra, V.V. and M. Govinda Rao (2003). "Trade Policy, Agricultural Growth and Rural Poor—Indian Experience, 1978-79 to 1999-2000", *Economic and Political Weekly*, 38(43), 4588-4603.

Samanta, R.K. (1994). "Issues for Agricultural Development in Tribal Dominated North East India", in *India's North East—The Process of Change and Development*, B.R. Publishing Corporation, Delhi.

Swaminathan, M. (1999). "Understanding the Cost of Food Corporation of India", *Economic and Political Weekly*, December 25.

7

CHANGING AGRICULTURAL SCENARIO IN NORTH-EAST INDIA IN THE ERA OF GLOBALISATION

— PRANAY JYOTI GOSWAMI

The agricultural practices of North-East India are of two types—(1) Hill (*Jhum*), and (2) Settled or plains agriculture. About three-fourths of her population, on the average, depend on agriculture and other allied activities. About 40 per cent of NSDP comes from agricultural sector.

The main objective of this chapter is to examine the changes that have taken place in agricultural sector in the North-East India with special reference to the following:

(1) Land-use pattern and operational holding;
(2) Hill agricultural scenario;
(3) Agricultural productivity, more particularly area and yield rate of different crops;
(4) Trend in input use with special reference to irrigation, HYV seeds and fertilisers;
(5) The impact of globalisation in the agricultural sector of North-East India.

The first-four aspects are analysed in four sections—Section I to Section IV. The conclusion and the impact of globalisation in the agricultural sector of North-East India is analysed in Section V.

The reference period of the present analysis is 1981-2001. Sikkim is excluded from our analysis.

SECTION I

Comparing the statistical data relating to 1982-83 and 1997-98, it is revealed that the forest area of Assam and Manipur has decreased at a very alarming rate whereas in case of Nagaland, the forest area has increased more than two and half times. In case of other States, the forest area has either remained the same or increased marginally. In case of net area, sown some improvement, though insignificant, has taken place in all States except Manipur (Table 7.1)

Area sown more than once has increased in Arunachal Pradesh, Assam, Manipur, Meghalaya and Tripura. But in case of Mizoram and Nagaland, it has decreased (Table 7.1).

During 1982-83 to 1997-98 the net irrigated area remains the same in Assam, Manipur, Mizoram. In Nagaland and Meghalaya the irrigated area has decreased. This is perhaps due to the failure of certain irrigational projects in this region (Table 7.2).

As regards operational holdings, marginal holdings in all States of North-Eastern Region are gradually increasing. This is also true for small and semi-medium holdings (Table 7.3). The average size of operational holdings (in hectares) in the States of North-East India except Meghalaya and Manipur is gradually decreasing (Table 7.4). This is perhaps due to rapid growth of population and lack of diversification of economic activities. Number of operational holdings, on the other hand, is gradually increasing (Table 7.3).

SECTION II

In the hills of North-East India, the majority of the tribal people practice shifting or *jhum* cultivation (Table 7.5). The method of cultivation is briefly described here.

The members of every family first select the place of hilly forest land where they want to grow their crops. They clear land covered by trees and undergrowths in the beginning of January. Then the place is allowed to dry for a few months. After that, they burn whatever remains in the surface of the land in the post-clearance stage. Just before the monsoon, the seeds are sown, broadly spread and dibbled. The ashes the land gets after burning becomes manures. The next phase is the phase of harvesting and storage. This phase begins in September and ends in November-December. As the

Table 7.1: Land Utilisation in North-East India

(In thousand hectares)

States	Geo-graphical area	Year	Reporting area for land utilisations	Forest area	Area put to non-agricul-tural uses	Barren & uncultiv-able land	Total of columns (6)+(7)	Net area sown	Area sown more than once	Total cropped area	Cultiv-able waste land*
(1)	(2)	(3)	(4)	(5)	(6)	(7)	(8)	(9)	(10)	(11)	(12)
Arunachal Pradesh	8,374	1982-83		5,154.0	NA	NA	NA	118	44	162	33
		1997-98	5,495	5,154.0	NA	48	48	185	65	250	
Assam	7,844	1982-83		3,070.8	NA	NA	NA	2,696	860	3,556	43
		1997-98	7,850	1,930.0	1,045	1,448	2,493	2,751	1,243	3,994	
Manipur	2,233	1982-83		1,515.4	NA	NA	NA	140	48	188	2
		1997-98	2,211	602.0	26	1,419	1,445	140	67	207	
Meghalaya	2,243	1982-83		851.0	NA	NA	NA	193	15	208	1
		1997-98	2,241	932.0	86	141	227	207	44	251	
Mizoram	2,108	1982-83		1,593.5	NA	NA	NA	65	13	78	—
		1997-98	2,109	1,598.0	62	65	62	109	4	113	
Nagaland	1,658	1982-83		288.2	NA	NA	NA	185	13	178	148
		1997-98	1,561	863.0	62	NA	62	251	9	260	
Tripura	1,049	1982-83		592.2	NA	NA	NA	246	135	381	8
		1997-98	1,049	606.0	133	NA	133	277	179	456	

* Relates to 1990-91.

Source: Basic Statistics relating to North-Eastern Region and CIME, different issues.

Table 7.2: Net Sown Area, Cropped Area and Irrigated Area of Different States of North-Eastern Region

(in '000 hectares)

	States	Years		
		1982-83	1991-92	1997-98
Net Area Sown	Arunachal Pradesh	118	149	185
	Assam	2696	2706	2751
	Manipur	140	140	140
	Meghalaya	193	202	216
	Mizoram	65	65	109
	Nagaland	165	190	220
	Tripura	246	270	277
	Total NER	3623	3722	3898
	All India	1,42,002	1,42,999	1,42,083
Total Cropped Area	Arunachal Pradesh	162	247	250
	Assam	3556	3797	3994
	Manipur	188	180	207
	Meghalaya	214	243	251
	Mizoram	65	74	113
	Nagaland	178	210	260
	Tripura	381	445	456
	Total NER	4744	5196	5531
	All India	1,65,800*	1,85,742	1,92,629
Net Irrigated Area	Arunachal Pradesh	24	31	36
	Assam	572	572	572
	Manipur	65	65	65
	Meghalaya	50	46	47
	Mizoram	8	8	8
	Nagaland	66	59	62
	Tripura	29	41	35
	Total NER	814	807	825
	All India	31,100*	53,510	54,563
Gross Irrigated Area	Arunachal Pradesh	24	32	36
	Assam	572	572	572
	Manipur	75	75	75
	Meghalaya	51	47	54
	Mizoram	8	8	10
	Nagaland	76	60	70
	Tripura	38	41	60
	Total NER	844	835	877
	All India	23,000*	63,204	72,784

* Relates to 1974-75.

Source: Basic Statistics of North-Eastern Region, Different issues.

Table 7.3: Number of Holdings by Size Groups in the States of NER

('000 Nos.)

States/ Holdings	Arunachal Pradesh	Assam	Manipur	Meghalaya	Mizoram	Nagaland	Sikkim	Tripura	West Bengal	All India
(1)	(2)	(3)	(4)	(5)	(6)	(7)	(8)	(9)	(10)	(11)
Marginal										
1985-86	14	1,451	67	59	21	8	13	211	4,343	56,147
1990-91	17	1,451	69	59	29	13	26	217	4,639	62,110
Small										
1985-86	16	546	48	51	19	19	10	70	1,175	17,922
1990-91	18	546	49	42	23	21	11	69	1,107	19,970
Semi-medium										
1985-86	27	324	21	46	11	19	9	27	516	13,252
1990-91	30	324	21	46	9	26	9	28	457	13,910
Medium										
1985-86	23	92	3	13	1	42	5	3	94	7,916
1990-91	26	92	3	13	1	47	5	4	79	7,630
Large										
1985-86	5	6	—	1	—	36	1	—	1	1,918
1990-91	4	6	0	1	0	33	1	0	1	1,670
All Sizes										
1985-86	85	2,419	140	171	52	125	38	312	6,130	97,155
1990-91	95	2,419	142	162	61	142	53	318	6,284	1,05,290

Source: CMIE, Report on Agriculture, November 1998.

Table 7.4: Average Size of Operational Holdings (in hectares) in the States of North East India

States	Years		
	1976-77	1985-86	1990-91
Arunachal Pradesh	NA	NA	3.62
Assam	1.37	1.31	1.31
Manipur	1.14	1.21	1.23
Meghalaya	1.67	1.67	1.77
Mizoram	NA	NA	1.38
Nagaland	7.90	7.15	6.82
Sikkim	2.67	2.50	1.89
Tripura	1.24	1.03	0.97
All India	2.00	1.69	1.57

Source: CMIE, Report on Agriculture, November 1997.

Table 7.5: Shifting Cultivation in North-East India

States	Area under shifting cultivation (sq.km.)	Fallow period (in years)	Minimum area of shifting cultivation, one form or other (sq.km.)	Number of families practising shifting cultivation
Arunachal Pradesh	700 (1.30)	3–10	2,100 (3.89)	54,000
Assam	696 (1.30)	2–10	1,392 (2.40)	58,000
Manipur	900 (1.29)	4–7	3,600 (5.14)	70,000
Meghalaya	530 (1.01)	5–7	2,650 (5.07)	52,290
Mizoram	630 (1.01)	3–4	1,890 (3.18)	50,000
Nagaland	190 (0.16)	5–8	1,913 (1.65)	1,16,046
Tripura	223 (0.52)	5–9	1,115 (2.59)	43,000
Total	3,869 (0.87)		14,660 (3.31)	4,43,336

Note: Figures in the parentheses indicate area available per family.
Source: Basic Statistics of North-Eastern Region.

peasants grow different crops like ginger, millet, cotton, chilly, etc. in the same plot, different crops are harvested in different periods. Digging sticks, khantis, spades, sickles, axes, *khurpis*, etc. are some of the instruments used in the *jhum* cultivation. Because of rapid growth of population and lack of other employment opportunities, annually 368.9 thousand hectares of land are brought under shifting cultivation (Task force of shifting cultivation, 1883). About five decades ago, the *Jhum* Cycle was 15-20 years which has declined to 4-5 years (Ramkrishna *et al.*, 1981).

It has been also revealed for the works of some researches (Roy, Niranjan) that the productivity as well as economic gain from shifting cultivation has been gradually decreasing.

It has been pointed out by many experts that "*Jhuming* has been exerting a negative impact on the tribal economy and hill ecology of preponderantly tribal belt". This system is responsible for soil-erosion, floods, climatic changes and destruction of many rare species of flora and fauna. Every year a big area of forest cover is lost due to shifting cultivation (Table 7.6).

Table 7.6: Change of Forest Cover (sq. km.)

States	Forest cover				Loss due to shifting cultivation		
	1991	1993	1995	1997	1991 to 1993	1993 to 1995	1995 to 1997
Arunachal Pradesh	68,757	68,661	68,621	68,602	70	NA	75
Assam	24,751	24,508	24,061	23,824	165	NA	257
Manipur	17,685	17,621	17,558	17,418	28	NA	603
Meghalaya	15,875	15,765	15,714	15,657	110	NA	75
Mizoram	18,853	18,697	18,576	18,775	256	NA	292
Nagaland	14,321	14,348	14,291	14,221	63	NA	573
Tripura	5,535	5,538	5,538	5,546	10	NA	—
Total	1,65,777	1,65,142	1,64,359	1,64,043	702	NA	1,875

Source: Basic Statistics of North-Eastern Region, different issues.

Seminar (18-19, June 1976) organised by NEICSSR, resolved that "shifting cultivation has to be replaced by an improved form of land management". Terracing, an alternative of *jhuming* has been evolved as a result of sincere research work of Indian Council of Agricultural Research, failed to attract even one per cent of about 33 lakh tribal population. This method is only popularised by the tribals of Arunachal Pradesh and among the Angamis.

SECTION III

Table 7.7 shows the yield rate of different crops in North-East India and India as a whole. The table reveals that yield rates of many crops in North-East India are much lower than India as a whole. However, the yield rates of small millet, groundnut, ginger, etc. are higher than India as a whole. Small millet is produced mostly in Arunachal Pradesh, Assam and Meghalaya. The yield rate in Arunachal Pradesh (941) and Meghalaya (746) is much higher than India as a whole (432). In Assam, the yield rate is (496) kg/hectare. Groundnuts in North-East India is mostly produced in Nagaland and Tripura. In case of ginger, the yield rate in Mizoram, Arunachal Pradesh and Meghalaya (around 6,500 kg/hectare) is much higher than the country as a whole (3,477). In case of gram, the yield are in NER is higher than India as a whole. Gram is mainly produced in Arunachal Pradesh, Meghalaya and Tripura.

If we examine the change of yield rate over time, it is clear that the yield rate of foodgrains in North-Eastern Region during 1980-83 was higher than India as a whole. But in later years, North-East India lagged behind the country as a whole. In case of wheat the gap was 674 kg/hectare in 1980-83, which has increased to 1,292 kg/hectares during 1996-97 to 2000-01, nearly double in two decades. In total pulses, the yield rate of NER is gradually improving and now there is no difference between North-East India and India as a whole. In case of mesta, the yield rate in North-East India as a whole upto 1991-92 was higher than India as a whole; but now yield rate in North-East India as a whole is lower than India as a whole. But within North-East, yield rate in Tripura (1,980 kg/hectare) is much higher than India as a whole (1,154.4 kg/hectare) in 2000-01 (Basic Statistics of NER, 2002).

However, (Table 7.8) reveals that the yield of some of the important crops in NER varies from State to State. In some State the yield rate of certain crops is higher than all India average whereas in some States, they lag behind.

As regards area under rice cultivation, like NER as a whole it is increasing in Nagaland, Mizoram, Assam during 1980-81 to 2000-01. In case of Tripura, there is falling trend. In Meghalaya during 1980-81 to 1990-91, the area has declined but during 1990-91 to

2000-01 it has increased. This is also true for Arunachal Pradesh. In Manipur during the first period, it has declined, then it slightly increased. For India as a whole, the area first increased and then decreased (Table 7.9).

Table 7.7: Yield Rate of Different Crops in NER and India as a Whole

(kg/hectare)

Crops	NER			India		
	1980-81 to 1982-83	1989-90 to 1991-92	1996-97 to 2000-01	1980-81 to 1982-83	1988-89 to 1991-92	1996-97 to 2000-01
Rice	1,114	1,240	1,474	1,291	1,688	1,896
Wheat	1,038	1,265	1,339	1,712	2,263	2,631
Maize	868	1,135	1,342	1,155	1,511	1,761
Small Millet	557	7,888	737	393	466	432
Total Pulses	374	527	647	491	553	580
Gram	—	539	737	618	466	432
Total Foodgrains	1,083	1,265	1,431	1,030	1,374	1,600
Ground Nut	—	1,245	1,066	813	917	969
Six Major Oil Seeds	415	420	577	578	753	858
Sugarcane	34,019	41,089	4,601	57,548	63,958	67,574
Jute	1,381	1,528	2,673	1,394	1,810	1,991
Mesta	922	1,024	996	826	979	1,068
Potato	6,411	7,318	8,912	13,244	15,938	19,002
Ginger	4,116	4,976	6,366	283	3,033	3,477
Chillies	—	487	932	638	819	1,062
Sweet Potato	3,286	3,928	2,869	7,339	8,236	8,422
Arecanut	952	929	881	855	1,148	1,193
Coconut	7,586	4,659	5,017	5,135	6,452	6,885

Note: Average production during 1980-81 to 1982-83, etc. is calculated by the author.

Yield of coconut is in terms of nuts per hectares.

Yield of Arecanut is in terms of kg. per thousand bearing plant.

Yield in other cases, kg/hectare.

Source: Basic Statistics of North-Eastern Region, different issues.

Table 7.8: Yield Rate of Different Crops in North-Eastern States

(*kg./hectare*)

Crops	States	Years		
		1980-81 to 1982-83	1989-90 to 1991-92	1996-97 to 2000-01
Rice	Arunachal Pradesh	1,049.0	1,162.0	1,085.0
	Assam	1,074.0	1,242.0	1,396.0
	Manipur	1,445.0	1,791.0	2,237.0
	Meghalaya	1,215.0	1,144.0	2,237.0
	Mizoram	1,908.0	1,212.0	1,781.0
	Nagaland	937.0	1,199.0	1,265.0
	Tripura	1,314.0	1,837.0	2,103.0
	Total NER	1,114.0	1,240.0	1,474.0
	All India	1,291.0	1,688.0	1,896.0
Wheat	Arunachal Pradesh	—	1,878.3	1,389.6
	Assam	1,145.6	1,214.3	1,286.8
	Manipur	—	—	—
	Meghalaya	—	1,300.0	1,653.3
	Mizoram	—	—	—
	Nagaland	—	3,200.0	3,111.2
	Tripura	—	1,859.3	1,966.6
	Total NER	1,038.0	1,265.0	1,339.0
	All India	1,712.0	2,263.0	2,630.6
Maize	Arunachal Pradesh	1,153.0	1,214.0	1,446.0
	Assam	601.3	631.6	732.3
	Manipur	1,720.5	2,626.6	3,348.3
	Meghalaya	1,046.3	1,247.3	1,455
	Mizoram	—	1,778.6	2,052.2
	Nagaland	700.3	936.6	1,112.3
	Tripura	—	—	1,000.0
	Total NER	868.0	1,135.0	1,342.0
	All India	1,155.3	1,511.0	1,761.0
Total Pulses	Arunachal Pradesh	—	—	1,000.0
	Assam	426.3	445.0	557.3
	Manipur	—	—	—
	Meghalaya	—	750.0	800.0
	Mizoram	—	883.5	2,000.0
	Nagaland	—	1,087.6	838.5
	Tripura	—	572.0	—
	Total NER	374.0	527.0	579.8
	All India	491.0	553.0	580.0

Table 7.9: Area under Different Crops

(in thousand hectares)

States	Crops	Years		
		1980-81 to 1982-83	1989-90 to 1991-92	1996-97 to 2000-01
Arunachal Pradesh	Rice	89.0	121.7	118.0
	Maize	23.2	37.6	33.5
	Wheat	—	3.5	4.0
	Total pulses	—	5.5	6.4
Assam	Rice	2,278.0	2,484.0	2,552.0
	Maize	20.6	19.6	19.4
	Wheat	103.2	84.5	80.8
	Total pulses	118.6	114.4	116.5
Manipur	Rice	172.0	160.0	161.6
	Maize	6.9	4.4	4.3
	Wheat	—	—	—
	Total pulses	4.5	—	—
Meghalaya	Rice	105.0	104.0	122.0
	Maize	177.0	18.5	17.0
	Wheat	—	4.6	4.2
	Total pulses	2.5	3.2	3.8
Mizoram	Rice	39.2	53.3	61.6
	Maize	4.2	6.3	7.9
	Wheat	—	—	—
	Total pulses	0.5	1.6	3.4
Nagaland	Rice	106.0	128.0	144.0
	Maize	17.7	23.9	31.0
	Wheat	—	1.6	4.5
	Total pulses	3.7	10.7	15.7
Tripura	Rice	292.0	260.4	252.5
	Maize	—	2.0	2.0
	Wheat	4.0	3.2	1.7
	Total pulses	5.3	10.9	9.7
Total NER	Rice	3,082.0	3,243.6	3,411.8
	Maize	90.4	111.8	115.6
	Wheat	113.7	96.1	93.0
	Total pulses	135.5	142.7	155.6
All India	Rice	39,707.3	42,183.1	40,070.7
	Maize	5,886.6	5,909.3	6,374.2
	Wheat	22,663.5	23,580.3	25,880.2
	Total pulses	23,044.4	23,451.4	21,773.3

Source: Basic Statistics of North-Eastern Region, different issues.

The total area of maize cultivation in Arunachal Pradesh during 1980-81 to 1990-91 has increased but during 1990-91 to 2000-01, it has decreased. In case of Assam, Manipur and Meghalaya the same trend is coming up. But in case of Mizoram, Nagaland the area of maize cultivation is gradually increasing. In Tripura, the area remains the same. However, the area of maize cultivation for the whole North-East India is increasing like India as a whole.

Area under wheat cultivation in Assam, Meghalaya and Tripura is gradually decreasing for the whole period under consideration. This is true for NER as a whole. But like India as a whole it is increasing in Nagaland and Arunachal Pradesh.

As regards area under total pulses, it is increasing for the country as a whole during the period 1980-81 to 1990-91, but it has declined during 1990-91 to 2000-01. But in case of NER as a whole, there is an increasing trend. The increasing trend is also true for the States like Nagaland, Mizoram, Meghalaya and Arunachal Pradesh. However, some exceptions are there in States like Assam and Tripura.

SECTION IV

The yield rate depends on a number of factors like consumption of fertilisers, use of HYV seeds, modern technique of production, availability of cheap credit facilities, etc.

Table 7.10 reveals that consumption of plant nutrients per unit of gross cropped area (kg/hectare) in all States of NER is much lower than India as a whole except Manipur in the year 2001-02. Among all the States of North-Eastern Region consumption of fertilisers is the highest in Manipur.

Net irrigated area remains the same in Assam, Manipur and Mizoram but fluctuating in some other States (Table 7.2).

There is an increasing trend of total cropped area (Table 7.2) during the period under consideration.

The per capita availability of bank credit is also very low in the States of NER as compared with India as a whole (Table 7.11).

The area under high yielding varieties in NER is also very low. There is no improvement in the use of HYV seeds in the post-reform period (Table 7.12).

Table 7.10: Consumption of Plant Nutrients per Unit of Gross Cropped Area

(*kg/hectare*)

Sl.No.	States	1984-85				1991-92				1996-97				2001-02			
		N	P_2O_5	K_2O	Total	N	P_2O_5	K_2O	Total	N	P_2O_5	K_2O	Total	N	P_2O_5	K_2O	Total
1.	Arunachal Pradesh	—	—	—	—	1.2	0.6	0.2	1.9	1.3	0.5	0.4	2.2	1.6	0.9	0.4	2.9
2.	Assam	2.1	0.8	1.0	3.9	5.1	2.1	2.2	9.5	8.4	1.5	4.7	14.6	19.2	10.6	9.1	38.8
3.	Manipur	17.4	2.9	0.3	20.6	33.4	13.2	1.3	48.0	42.0	4.6	2.0	48.6	86.5	12.0	6.4	104.0
4.	Meghalaya	7.9	5.5	0.7	13.6	6.4	6.1	1.4	13.8	9.2	4.4	0.8	14.4	10.5	6.2	0.5	17.2
5.	Mizoram	—	—	—	—	5.0	5.5	0.5	11.0	2.1	0.9	0.5	3.5	4.9	5.2	3.6	13.7
6.	Nagaland	0.9	0.6	—	1.5	1.4	1.8	0.4	3.5	1.9	1.6	0.4	3.8	1.1	0.8	0.2	2.1
7.	Tripura	5.5	1.5	1.3	8.3	1.58	6.2	3.1	25.0	11.8	4.1	2.9	18.8	1.5	8.6	5.3	30.5
	All India	30.4	10.4	4.6	45.5	44.4	18.3	7.5	70.3	54.8	15.8	5.5	76.0	58.7	22.8	8.7	90.1

Source: Basic Statistics of NER, various NEC, Crop, Shillong issues.

Table 7.11: Credit to Agriculture in the States of NER

(*Rs. per capita*)

States	1980	1985	1990	1995	2000 (Oct.)	CARG 1980	CARG 1990
Arunachal Pradesh	1	9	54	112	154	43.94	12.40
Assam	29	41	119	136	108	15.12	–1.09
Manipur	8	17	54	70	104	21.59	7.54
Meghalaya	7	38	92	278	121	29.76	3.16
Mizoram	NA	23	113	188	95	83.80	–1.96
Nagaland	4	44	96	200	126	35.83	3.10
Sikkim	NA	29	140	145	160	36.90	1.45
Tripura	21	37	153	175	207	21.97	3.37
West Bengal	29	43	104	105	129	13.58	2.42
All India	NA	118	201	271	408	11.15	8.21

Source: CMIE, Profiles of Districts, October 2000.

SECTION V

From the above analysis it is seen that there is stagnation in the state of agricultural development in the NER even in the post-reform period. In terms of use of different types of inputs, this region still lags behind India as a whole.

In India during 1980s we noticed a tendency of openness of the economy. In 1991, Government of India announced some economic reform measures in order to promote competitiveness in our economy. In the meantime, WTO was established on 1st January, 1995 and India became the founder member of WTO by ratifying in WTO agreement on 30th December, 1994. The basic objectives of WTO Agreement on Agriculture (AoA) were to initiate substantial progressive reduction in support and protection resulting in fundamental reforms in the world agricultural trade and to make it more transparent. There are three commitments mandated in the AoA, i.e., market access, domestic support and export competition. Again competitiveness in the global market is multidimensional concept. It, involves not only price competitiveness but also the ability to deliver the contractual quality consistently at the appropriate time and place.

Table 7.12: Area Under High Yielding Varieties in North-Eastern Region

(*in '000 hectares*)

States		Paddy			Wheat			Maize		
		1986-87	1991-92	1996-97	1986-87	1991-92	1996-97	1986-87	1991-92	1996-97
Arunachal Pradesh	T.A.	116.0	115.3	115.8	2.1	4.6	4.3	30.0	37.6	31.2
	H.Y.V.	6	22	35	0.7	4	5*	2	14	—
Assam	T.A.	2,287	2,520.6	2,453.5	120.0	73.9	87.9	18.4	20.6	19.1
	H.Y.V.	1,030	1,150	1,422	121*	76*	103	7	—	12
Manipur	T.A.	165.1	157.4	166.1	—	—	—	7.5	4.6	6.4
	H.Y.V.	83	90	72	8	0.5	0.3	5	—	7*
Meghalaya	T.A.	108.5	103.6	104.8	4.8	4.6	4.3	19.2	18.4	17.0
	H.Y.V.	29	37	42	6*	4	6*	10	10	15
Mizoram	T.A.	48.3	57.3	64.7	—	—	2.0	4.0	6.6	8.5
	H.Y.V.	3	10	5	—	0.03	—	4	4	—
Nagaland	T.A.	121.1	128.4	140.0	—	—	2.0	22.0	23.9	32.0
	H.Y.V.	18	30	2	—	0.4	8	5	9	3
Tripura	T.A.	256.9	274.0	258.9	2.7	3.7	2.0	—	—	2.0
	H.Y.V.	152	207	223	3*	4*	2	—	2	2
All India	T.A.	41,166.7	42,596.3	43,433	23,131.2	23,977.2	25,887.1	5,923.1	5,953.6	6,260.4
	H.Y.V.	24,026	27,394	33,399	19,192	20,968	23,726	2,185	2,608	3,764

* Doubts are there regarding the authenticity of data because area under HYV seeds is more than total area (TA) under cultivation.

Source: Basic Statistics of North-Eastern Region, 1992-2000.

So, in these days of globalisation, we have to compete with other countries in price front, quality front and service front. So, Government must come forward to modernise/develop power, telecommunication, roads, airports, etc. One of the demerits of globalisation is commercialisation of agriculture. As a result, there is every chance of loss of biodiversity in our country. In this respect, Government should take appropriate measures so that biodiversity of North-Eastern Region may be preserved.

Again, there is much potentialities of floriculture, horticulture, pisciculture in this region. There is also a good prospect for cultivation of medicinal herbs in this region. One of the medicinal plants called 'Texas Buccata' available in Arunachal Pradesh is well known for its use as a medicine for treatment of cancer. This is one of the costliest plants in the world and has a tremendous commercial value.

The environmentalists warn us regarding excessive and non-judicious use of plant protection chemicals. This has not only resulted in environmental pollution but also developed resistance to several pests. Residues of DDT, aldrine, etc. have been found in different food samples.

North-Eastern Region has been identified as one of the 25 hot-spot zones of the world from the point of view of biodiversity. In North-East India, which covers a large area by forest, there is every possibility for the manufacture of organic inputs. Having taken cognisance of the hazards associated with pesticides, emphasis has been given on biological pest management, which is being done either through cultural, biological or organically accepted chemical alternatives.

The demand for organic food is also increasing in foreign countries. North-Eastern Region, being a low chemical fertiliser consuming area has a good opportunity to take a production of organic food for export and domestic use. ICAR should come forward to invest organic pesticides, fertilisers, etc. by utilising the local resources.

There is also much scope to increase the cultivable area as there is much cultivable waste land in North-Eastern Region (Table 7.1) and most of the land is single cropped. By providing irrigation and other facilities, it is also possible to increase the gross cropped area in this region.

NOTE

1. At present Sikkim is also included in North-Eastern Council. Including Sikkim total number of North-Eastern States is 8. The other States are Assam, Meghalaya, Mizoram, Nagaland, Tripura, Arunachal Pradesh and Manipur.

   After Independence present Arunachal Pradesh, i.e., the then NEFA remained a part of Assam till becoming the Union Territory in 1972. State of Nagaland came into being in 1963; State of Meghalaya and Union Territory of Mizoram were also carved out of Assam in 1972. Arunachal Pradesh and Mizoram were upgraded to full Statehood in 1986.

REFERENCES

Bagchi, Kanak Kanti, "Relative Levels of Agricultural Development in the North-Eastern States—Current Status and the Imperatives Under Economic Reforms", in Sen Gupta Keya and Roy Niranjan (ed.), *Economic Reforms and Agriculture Development in North-East India*, Mittal Publication, New Delhi, 2003.

Bezbaruah, M.P., "A Review of Economic Trends in India's North-Eastern Region Since Independence", *Journal of North-East India Council for Social Science Research*, Shillong, April, 2001.

Choudhury, Rabindra Kumar, "Agricultural Economic Development in North-East: Chief Areas of Constraint and Remedies", *Journal of North-East India Council for Social Science Research*, October, 1996, p. 51.

CSO, Government of India, *Statistical Abstract*, India, 2001.

Dubey, Amaresh, Das Smita and Pala Veronica, *Migration in the North-Eastern Region of India: Some Issues*, The Paper was presented in a Seminar on Globalisation at NEHU, Shillong on 22-23 November, 2002.

EPW Research Foundation, *National Accounts Statistics of India, 1950-51 to 1996-97*, Published in 1998.

Goswami, Pranay Jyoti, "Agriculture Development in North-East India with Special Reference to Settled Agriculture" in Alam, K. (ed.), *Agricultural Development in North-East India—Constraints and Prospects*, Deep & Deep Publication, New Delhi, 1993.

Goswami, Pranay Jyoti, "Impact of Economic Reform on Agricultural Development in North-East India with Special Reference to Assam", in Sen Gupta Keya and Roy Niranjan (ed.), *Economic Reforms and Agriculture Development in North-East India*, Mittal Publication, New Delhi, 2003.

Government of India, *Census Reports*, 1951 to 2001, Various issues.

NEC, *Basic Statistics of North-Eastern Region*, 1987, 1992, 1995 and 2000.

NSSO, Government of India, *Sarvekshna*, Different issues.

Panda, Sudhakar, *Globalisation and the Challenge of Poverty, Unemployment and Social Development in North-Eastern States*, the Paper was presented in a Seminar on Globalisation in NEHU, Shillong on 22-23 November, 2002.

Registrar General India, New Delhi, *Vital Statistics of India*, 1993 and 1994.

Roy Burman, B.K., "Land, People and Demographic Indicator of North-East India", *Journal of North-East India Council for Social Science Research*, December, 1999, p. 63.

8

AGRICULTURE IN THE TWENTY-FIRST CENTURY IN THE NORTH-EASTERN STATES
Challenges and Opportunities

— AMALESH BANERJEE

Introduction

Even after fifty years of planning, agriculture still remains the main focus of our development strategy. The Tenth Five Year Plan emphasises that even though the wage goods problem has been partly met with the overall growth of agricultural output in aggregate, agriculture still remains as the major driving force of the Indian economy. Share of agriculture in GDP has come down from 44 per cent in 1973-74 to 27 per cent in 1999-2000. In the North-Eastern States the importance of agriculture is crucial to development. Employment, poverty index and the SDP growth of these States indicate that agriculture occupies the largest share of employment and the strategy of agricultural growth in order to reduce poverty and increase employment is all the more important in the North-Eastern States. The important fact remains that notwithstanding Green Revolution agricultural development remains lopsided—areas-specific and crop-specific. In recent years with the growth of population per capita availability of foodgrains has been declining and food insecurity has become an endemic feature.

Seven contiguous States of North-East Region covers an area of 2,55,000 sq. kms. representing around 7.7 per cent of total geographical area of the country. The region has wide variations in physical features ranging from the plains of the Brahmaputra in

Assam to dissected mountains of eastern Himalayas in Arunachal Pradesh. It can be classified into three zones: (a) the North-Eastern hills and basins which accounts for 65 per cent of the total area, (b) the Brahmaputra Valley which covers 22 per cent of the region, and (c) the Meghalaya plateau covering 13 per cent of the total area.

The North-East Region supports a population of 38.4 million, around 4 per cent of country's total population and Assam itself account for 70 per cent 26.6 million of the region's population. The decadal growth rate during the period 1991-2001 has been at 21.9 per cent which is slightly higher than national average of 21.34 per cent. Except in Tripura and Assam the population growth of other individual States has been above the national average of 21.34 per cent. Nagaland the highest (64.41%) decadal growth of population in the country.

Recent growth performance of these States has not been encouraging. Except Tripura, which showed a rate higher than national average, all other States of the region had lower growth rate during the period 1993-94 to 1999-2000. The Tenth Plan (2002-07) envisages GSDP average annual percentage of growth rate for Assam 6.17, Manipur 6.46, Meghalaya 6.30, Mizoram 5.29, Nagaland 5.56, Arunachal Pradesh 6.84, Tripura 7.31. The agriculture is assumed to grow at 3-4 per cent per annum over this period in all these States except Mizoram which is expected to achieve only 2 per cent. The performance of NE States except Tripura thus appears to remain below Rajasthan, M.P., U.P. and Orissa which are supposed to be lagging in the process of development. One can compare this with West Bengal (5.09), most shining State in respect of agricultural performance at present. It is obvious that higher performance in agriculture contributes to the reduction of poverty. In West Bengal over the last decade agricultural growth has largely contributed in reduction of poverty. Per capita net State domestic product has been very low in Assam, Nagaland, Arunachal Pradesh; the exponential growth rate of per capita NSDP over the period, 1993-94 to 1999-2000, has been 0.4 for Arunachal Pradesh and Assam and less than one for Nagaland. However, other NE States have better record. Exponential growth rate of per capita net SDP was 4.8 for Manipur, 4.5 for Tripura, 2.8 for Meghalaya. Compared to that of West Bengal has 5.7, Karnataka 5.8, Kerala 3.6, U.P. 3.2, Bihar 2.7, Orissa 2.0 (Tenth Plan, Vol. III, Ch. 3, Annexure 3.1, 3.2). The purpose of this presentation is to highlight the importance of agriculture in reducing poverty, reducing per capita income and employment. In this chapter we examine the nature of agriculture in the NE States and the growth model fitting in this area.

Table 8.1: Poverty Ratio in 1973-74 and 1999-2000

N.E. States	1973-74	1999-2000
Assam	51.21	36.09
Mizoram	50.32	19.47
Manipur	49.96	28.54
Nagaland	50.81	32.67
Arunachal Pradesh	51.93	33.47
Meghalaya	50.20	33.87
Tripura	51.00	34.44
Sikkim	50.86	36.55
All India	54.88	26.10

Source: Tenth Plan.

Table 8.2: Poverty Reduction and Growth Targets: Potential of the States in the Tenth Plan

	State-wise Growth Target (Per cent)			GSDP Growth (Per cent)	Poverty Reduction Targets for 2007 (Percentage)		
	Agri-culture	Industry	Service		Rural	Urban	Com-bined
Arunachal Pradesh	4.00	8.90	10.50	8.05	37.89	4.48	29.33
Assam	3.82	5.00	9.00	6.17	37.89	4.48	33.33
Manipur	3.59	8.33	7.39	6.46	37.89	4.48	30.52
Meghalaya	4.00	6.87	7.05	6.30	37.89	4.48	31.14
Mizoram	2.00	4.16	6.84	5.29	37.89	4.48	20.76
Nagaland	4.00	7.29	5.78	5.56	37.89	4.48	31.86
Tripura	3.90	9.37	8.43	7.21	37.89	4.48	31.88
Sikkim	5.00	5.21	10.36	7.87	37.89	4.48	33.76
West Bengal	5.09	9.15	10.76	8.75	21.98	8.98	18.30
Uttar Pradesh	4.67	11.05	7.92	7.61	24.25	26.17	24.67
Kerala	3.05	5.89	8.17	6.46	1.63	9.34	3.61
All India	4.00	8.86	9.35	8.00	21.07	15.06	19.34

Source: Tenth Five Year Plan, Vol. III, Tables 6.5 and 6.6.

Table 8.3: Comparative Position of Per Capita Flows to States 1999/2000/2001—NE and Leading States

States	Population	Per Capita NSDP (Rs.)
Andhra Pradesh	75,727,541	14,715
Bihar	82,878,796	6,328
Gujarat	50,596,992	18,685
Haryana	21,082,989	21,551
Karnataka	32,733,958	16,343
Kerala	31,838,619	18,262
M.P.	60,385,118	10,907
Maharashtra	96,752,247	23,398
Orissa	36,706,920	9,162
Punjab	24,289,296	23,040
Rajasthan	56,473,122	12,533
Tamil Nadu	62,110,839	19,141
Uttar Pradesh	166,052,859	99,765
West Bengal	80,221,171	15,569
Arunachal Pradesh	1,091,117	14,338
Assam	26,638,407	9,720
Manipur	2,388,634	11,370
Meghalaya	2,306,069	11,678
Mizoram	891,058	N.A.
Nagaland	1,988,636	18,262
Sikkim	540,493	13,356
Tripura	3,191,168	10,213

Source: Tenth Plan, Vol. III, Table 3.23.

Growth Models and North-East

As we have seen above, agricultural growth pattern in different States is at different levels. This is obvious due to historical and socio-economic condition of States. Although the land ceiling act has been passed in some States but the act has not been enacted in right earnest due to the existence of stronghold of vested interest of the feudal class. Besides a sizeable section of population comprising tribals and other marginal section of population are traditionally landless or have no ownership on land. Unless land reform measures are implemented these large section of the poor people will remain weak and dependent on the richer landholding section for food, credit and security.

Conceptionally we can conceive of two types of growth models based on agriculture. One is the surplus accumulation through agricultural development by application of modern techniques in large farms where landless peasants work as farm labourers. At a certain stage of our planning the Green Revolution has emerged basically from this type of historical structure. There has been substantial accumulation from green revolution. Output increased and the public authority supported the process through different types of subsidies. Both the farm owners and labourers have improved their positions. However, this high-tech Green Revolution remained area-specific and crop-specific. Large parts of the country, although without land reform, could not adopt the Green Revolution technique. Particularly the NE States remained outside that system of cultivation. The second mode of cultivation is based on land reform. Under this technique land reform is implemented and the share croppers and small farmers have been given land ownership. Karnataka, Kerala and West Bengal have implemented this strategy and with the support of modern inputs the farmers have increased the productivity of their existing land. West Bengal, with this structural reform has brought about significant improvement in farm output. The Tenth Plan estimated that land reform had contributed to the highest agricultural growth of West Bengal. Both these models have wide socio-economic ramifications.

Agriculture in the NE States does not fall in any of these two models. Although endowed with abundant water the NE States could not adopt the high-tech farming. Secondly since the land reform programme has not been implemented, West Bengal type of farming also has not developed. Agriculture in the NE States has a number of special characteristics such as hill terrain, community ownership of land among the tribal community, shifting cultivation and regular devastating flood every year. Land improvement measures such as application of modern techniques and inputs are not present. In the absence of improvement of productivity of land and existence of growing unemployment, the pressure on land increases. Rate of agricultural growth can hardly meet the demand of growing population. The economy of NE States is almost trapped in a state of vicious circle of poverty and low level equilibrium.

Jhum **Cultivation and Environment**

Jhum cultivation is the primitive type of family or community cultivation on the hill slope. Hill sides are prepared for cultivation by slash and burn of bush for a short period. Family labour and handtools are the inputs. Tribals have traditionally practiced this type of cultivation on the hills for their subsistence. Human beings without support of animal or plough make the hill sites arrable and produce for their limited needs with the support of natural rains. Such cultivation, therefore, by nature produce no surplus as the slope of the hill suffers erosion and depletion of fertility after sometime. Jhumias then move on to other slopes and repeat the same process.

Recommendations

Due to logistic problem the North-Eastern Region may be considered, for all practical purposes, a closed economy separated from the rest of the country. As mentioned above, the peculiarities of agriculture land support to this proposition:

(a) In the new regime of liberalisation, agriculture faces many problems. According to W.T.O. rules which India, as a participant of W.T.O. has to follow, foreign agricultural products are allowed to come in our country and our products are restricted by tariff wall in those countries. The impact of this will also affect prospect of agriculture in NE States. Horticultural and other agro products which are exported in foreign countries will face fierce competition.

(b) In our closed economy model of NE agriculture has to be developed in those fields where the factor endowment is rich. As stated above, NE has wide field of forest cover which facilitates extensive cultivation of horticultural products. The main point is that if agriculture has to develop according to comparative advantage those products have to be marketed. There is enough of marketable surplus but not sufficient marketed surplus.

(c) Rubber plantation, mushroom, fishery, dairy, processing of fruits and other agro-products are some of the products of diversified agriculture. In that way traditional agriculture will have to be modernised. *Jhum* cultivation has no romantic appeal. It is simply primitive way of living. Settled cultivation of the tribals alongwith complementary activities will transform their way of life in a more modern way.

(d) Food security and poverty eradication remains the main objective of agricultural development. As stated above, the NE States have a poverty level at the same rate of national level. Raising income from agriculture should be the major task of agricultural transformation. The NE should have a separate food policy based on the nature of production and pool of foodgrains produced in these States.

(e) Land reform and application of modern inputs will raise agricultural output, as it happened in West Bengal and consequently helped substantially to reduce food insecurity and poverty. As mentioned above, institutional reform and empowerment of farmers with assets and institutional support will go a long way in raising productivity and income of the poor peasantry. This will help increasing the effective demand and employment in the rural sector.

(f) Water management is a serious issue all over India. In the North-Eastern States as also in the entire Himalayan foot-hills excess supply of water flowing from hills cause devastation. Even after fifty years no serious effort has been made to control the flood of Assam and North Bihar. The NE States have great potentiality of power generation which can be utilised for further development of the area.

(g) In the tribal economy there is the phenomenon of non-market production as against commercial or market production. This non-market farming of the tribal hamlet is called the Z-goods model of agricultural households. The main feature of such households is that they produce for their own consumption in their small plots. They supplement the other requirements by grazing, water carrying, handicrafts. The basic commodities are here termed as Z-goods. The households in this secluded set-up maximise their utility by spending time in home products and supplementing that with market goods. It is a tremendous job to raise their desire for higher level of living and to bring their productive effort in the market channel.

Conclusions

Sustainable development of agriculture must pay special attention to the ecological balance. Millions of households of tribal population locked in untractable terrain do have virtually no knowledge of modern cultivation. Innovative techniques have to be applied to train them and bring them to the market economy from the trap of vicious circle of poverty and malnutrition. Physical and financial infrastructure have to be strengthened. On the one hand increasing output, storage and marketing and distribution have to be strengthened. Simultaneously, investment and credit availability have to be ensured. Growth of North-Eastern economy requires massive exogenous flow of capital from the central government for water management and infrastructure development which will have multiplier effect on employment and income generation.

REFERENCES

Basu, Kaushik, *Agrarian Question*, O.U.P., 1994, New York.

Chakraborty, Sukhamoy, *Development Planning*, 1987, Oxford, New York.

Colman, David and Trevor Young, *Principles of Agricultural Economics*, Cambridge University Press, Cambridge, 1989, New York.

Government of India, *Tenth Five Year Plan*, 2003, New Delhi.

Karmakar, K.G., *Rural Credit and Self-help Groups*, Sage Publications, 1999, New Delhi.

9

PROSPECTS OF DRIP IRRIGATION IN NORTH-EASTERN HILL REGION

— R.K. SINGH, K.N. AGRAWAL and K.K. SATAPATHY

Fresh water is a finite and vulnerable resource, essential to sustain life, development and environment. The amount of water readily available for human development is less than 0.01 per cent of the total water available in the hydrosphere (i.e., 1.4 billion km^3), which is renewed annually. For adequate socio-economic conditions in a country, available renewable water resources necessary has been estimated to be 1000 m^3/person/year. The per capita availability of water resources in India is estimated as 1,580 m^3/year which may shrink to 1,190 m^3/year in the year 2010 and 810 m^3/year in the year 2025. Without water security, one cannot expect food security for the growing population. The primary use of fresh water is irrigation, domestic use, industrial use, and hydroelectric power generation. Apparently irrigation is the biggest consumer of water in India, accounting for about 83 per cent of total water usages. Mostly in our country, the conventional mode of irrigation such as flood irrigation, furrow irrigation, border irrigation, check basin irrigation, etc. is practiced. In this method water is replenished to the field capacity level after depletion by 50 to 60 per cent of the field capacity. The water use efficiency of these methods is very low ranging from 25 to 40 per cent. The field is either over irrigated at the head end or under irrigated at the tail end. Waterlogging problems may also occur. These conditions have created the low productivity levels of 2-3 tonnes/ha in irrigated agriculture against 4-6 tonnes/ha at research farms. The use of drip or trickle irrigation system may solve the problem faced by the conventional mode of irrigation and also contribute significantly towards saving the precious water resources. In the North-Eastern Hill (NEH) region where production and productivity of crops are low due to water stress during the dry period, drip irrigation may play a vital role. Drip irrigation is frequent application of required and measured quantities of water directly on or below the soil surface near the root zone of plant usually as discrete drops through emitters placed along a low pressure delivery system consisting of piping network and emitting devices. The system encompasses a number of methods or concept such as Inline Drip, Online Drip, etc. depending upon the emitting device characteristics, and functions. A standard Drip Irrigation system consists of physical components required to deliver water to the root zone of crop in form of fine droplets, viz., pump unit, filtration unit, fertilizer application, main pipeline, sub-main pipeline, lateral and emitting devices (drippers/dripline), valves and other accessories as needed. The optimum air and moisture level in the soil and the root zone of the plant is maintained by this method of water application throughout the growing period resulting in proper uptake of the nutrients. Thus Drip Irrigated crops give higher yields with less quantity of water. Drip Irrigation can be adopted to apply water in more than 80 crops including sugarcane, fruits and vegetables, fibres, nuts, oilseeds, orchards, plantation crops, spices and cotton.

Benefits of Drip Irrigation

Water is applied at frequent intervals in drip irrigation and hence the growth of the crop is uniform. The crop growth is not affected due to water/moisture stress. Due to application of measured quantity of water to each plant based on its evapotranspiration requirements, the overall irrigation water requirement is lowest.

Water Use Efficiency: Application of water in accurate quantity and at desired location reduces evaporation losses and eliminates overspray. With proper design and scheduling, water is used only when and where it is needed. Water use efficiency of more than 90 per cent can be achieved through drip irrigation. This saves water up to 60 per cent as compared to the conventional method of irrigation.

Optimum Growing Conditions: For healthy and vigorous growth of the plants, an optimum balance of oxygen and moisture around their root zone is to be maintained. Slow application rates maintain this optimum condition, and proper scheduling will eliminate a wet/dry fluctuation that stresses plants. Slow application of water also minimises erosion, leaching of nutrients beyond root zone, and compaction of the soil.

Reduced Pest Problems: As the plant foliage remains dry, it prevents water-borne pests, especially fungal diseases and leaf burns that are sometimes evident in other modes of irrigation such as sprinkler irrigation. Moreover, in orchards, plantations and vineyards the area between the wetted strips is kept dry. This facilitates the movement of men and machinery while irrigation is in progress.

Reduced Weed Problems: Since the water is applied to root zone only, the remaining area of the field remains dry. This reduces the weed growth and competition for water and nutrients remains at low between desired and undesired plant population.

Versatility: Drip irrigation can be used in more than 80 crops. With the introduction of pressure compensating drippers, uniform delivery of water is ensured on undulating landscape. Even mild saline water can be used for irrigation by this method.

Quality of Produce: Big and uniform size of produce is obtained owing to the continuous growing conditions available to the plants. Quality of the produce is also much better due to reduction in the incidence of insects, pests and diseases.

Reduced Labour: Undulating terrains, rolling topography and hilly slopes can be easily irrigated without making investment on land levelling and shaping.

Water saving and yield obtained by use of drip irrigation in different crops are presented in Table 9.1.

Fertigation with Drip Irrigation

Concept of fertigation was evolved in the late 1970s to serve the farming community. Fertigation (fertiliser plus irrigation) is an effective method of applying chemicals and fertilisers to crops using modern irrigation techniques. Usually nitrogen, potassium sulfate and micronutrients such as iron, zinc, copper and manganese can be injected as chlorates or sulfates using drip irrigation system. By the process of fertigation, irrigation water and fertilizer in liquid form is placed at the root zone of the crop. Thus higher water as well as fertiliser use efficiency is achieved. Fertigation reduces fertiliser

Table 9.1: Water Saving and Yield Enhancement due to Drip Irrigation

Sl.No.	Crop	Yield (Q/ha)		Irrigation (cm)		WUE (q/ha/cm)		Adv. of MI (%)	
		SI	Drip	SI	Drip	SI	Drip	Water Saving	Inc. in Yield
1.	Beet	5.70	8.90	86.00	18.00	0.07	0.50	79.10	56.10
2.	Bitter guard	32.00	43.00	76.00	33.00	0.42	1.30	56.60	34.40
3.	Brinjal	91.00	148.00	168.00	64.00	0.55	2.30	61.90	62.60
4.	Broccoli	140.00	195.00	70.00	60.00	2.00	3.25	14.30	39.30
5.	Cauliflower	171.00	274.00	27.00	18.00	63.00	15.20	33.30	60.20
6.	Chilly	42.30	60.90	109.00	41.70	0.39	1.50	61.70	44.00
7.	Cucumber	155.00	225.00	54.00	24.00	2.90	9.40	55.60	45.20
8.	Lady-finger	100.00	113.10	53.50	8.60	1.87	13.20	84.00	13.10
9.	Onion	284.00	342.00	52.00	26.00	5.50	13.20	50.00	20.40
10.	Potato	172.00	291.00	60.00	27.50	2.90	10.60	54.20	69.20
11.	Radish	10.50	11.90	46.00	11.00	0.23	1.10	76.10	13.30
12.	Sweet potato	42.40	58.90	63.00	25.00	0.67	2.40	60.30	38.90
13.	Tomato	61.80	88.70	49.80	10.70	1.24	8.28	78.50	43.50
14.	Banana	75.00	875.00	176.00	97.00	3.27	9.00	45.00	52.20
15.	Grapes	264.00	325.00	53.00	28.00	5.00	11.60	47.20	23.10
16.	Papaya	130.00	230.00	228.00	73.00	0.60	3.20	67.90	76.90
17.	Pomegranate	34.00	67.00	21.00	16.00	1.62	4.20	23.80	97.00
18.	Watermelon	82.10	504.00	72.00	25.00	5.90	20.20	65.30	513.90

consumption by 30-50 per cent compared to the conventional methods of fertiliser application. Fertigation is possible through tank, ventury or pump systems. The most practical method of applying of fertilisers to the irrigation system is by creating a 10 per cent bypass flow of the main line flow, through an artificial fertiliser mixing tank. Experiments conducted at Indian Agricultural Research Institute on fertigation of broccoli indicated that substantial savings of fertiliser (20 to 40%) could be accomplished through this method. Most of the grape growers of Maharashtra have adopted fertigation for grapes.

Coverage of Area under Drip Irrigation

Considering the advantages of the drip irrigation, particularly in saving the precious water resources, Government of India had given subsidy till 1999-2000. Due to this policy area under drip irrigation increased from 1,500 ha in 1985 to 3 lakh ha with an average growth rate of 30,000 ha per year.

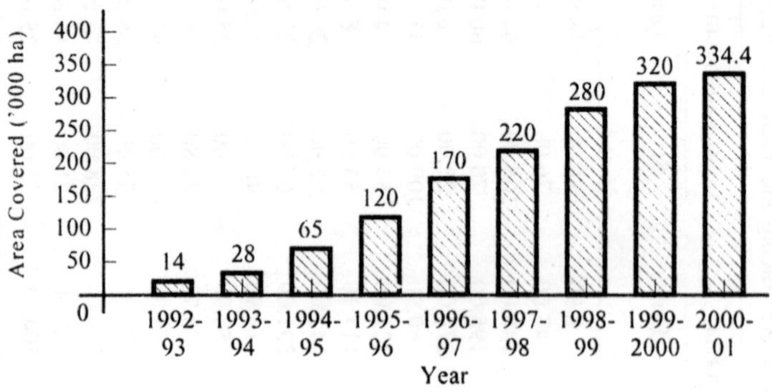

Fig. 9.1: Trends of Growth of Drip Irrigation

As per the records of the Government of India schemes, so far 1,592 ha has been brought under drip irrigation in the NEH region as shown in Table 9.2.

For the farmers of Meghalaya the concept of the drip irrigation is not new. Farmers of the East Khasi Hills and Jaintia Hills are practicing "Bamboo Drip Irrigation" for the irrigation of the crops like beetle vine, pepper and arecanut for last two centuries. This is a very useful irrigation system in the area where water is scarce, soils have poor water holding capacity and the topography is rocky and undulating. The system is based on the principle of the gravity flow.

Table 9.2: Area Covered under Drip Irrigation in NEH Region

Sl.No.	States	Area Covered (ha)
1.	Arunachal Pradesh	295
2.	Manipur	341
3.	Meghalaya	16
4.	Mizoram	124
5.	Nagaland	816
6.	Tripura	0
	Total	1,592

There is a vast potential for the drip irrigation in the NEH region to boost the production of horticultural crops in the region. The area under the horticultural crops and their production as compared to the national figure in the year 1999-2000 is presented in Table 9.3.

Table 9.3: Area and Production of Horticultural Crops (1999-2000)
(*Area in M ha and production in M tonnes*)

Sl.No.	Crop	All India		North-East	
		Area	Production	Area	Production
1.	Fruits	3.78	45.50	0.61	2.00
2.	Vegetables	5.99	90.83	0.37	4.05
3.	Spices	2.52	2.91	—	—
4.	Coconut	1.78	8.42	0.02	0.12
5.	Cashew	0.69	0.52	—	—
6.	Others	2.78	1.75	—	—
	Total	17.54	149.93	1.00	6.17

Table 9.4 shows the state-wise area and production of major horticultural crops in the region.

There is vast potential to the tune of 1 million ha for the drip irrigation in the region. As compared to this the area covered so far is negligible.

Constraints in Adoption of Drip Irrigation

In spite of the technical and economic feasibility, its large scale and fast adoption is not so encouraging in India. The following factors were identified in an all India survey of farmers:

Table 9.4: State-wise Area and Production of Major Horticultural Crops in NEH Region

(Area in thousand hectares, Production in thousand tonnes)

Fruit Crops	Arunachal Pradesh		Manipur		Meghalaya		Mizoram		Nagaland		Tripura	
	Area	Prod.	Area	Prod.	Area	Prod.	Area	Prod.	Area	Prod.	Area	Prod.
Pineapple	6.63	29.92	9.95	69.65	8.79	77.20	1.03	7.15	18.10	44.71	5.03	44.47
Orange	5.57	11.84	0.83	3.32	6.96	32.98	7.01	28.07	1.26	18.03	4.48	21.00
Arecanut	—	—	—	—	9.50	12.10	0.70	0.10	—	—	1.80	3.50
Banana	2.65	4.49	1.41	10.72	5.37	63.70	2.81	14.78	1.51	23.36	4.93	33.63
Mango	0.07	0.064	0.16	0.88	—	—	0.48	2.56	0.17	0.46	5.02	41.27
Papaya	0.46	1.95	1.85	10.17	0.48	3.92	0.296	2.42	0.285	2.48	—	—
Jack-fruit	0.12	0.71	0.26	1.04	—	—	0.22	1.39	0.225	0.91	8.13	214.80
Apple	2.37	5.41	—	—	—	—	0.032	0.031	0.083	0.45	—	—
Pears	0.58	1.67	—	—	—	—	0.093	0.44	—	—	—	—
Plum & Peach	0.73	1.89	0.61	3.97	—	—	0.18	0.52	0.93	3.43	—	—
Potato	4.50	33.30	3.20	20.80	20.30	200.50	0.70	3.70	3.80	30.70	4.90	91.30
Cabbage	—	—	0.95	1.30	1.00	2.50	NA	NA	NA	NA	NA	NA
Sweet Potato	—	—	0.04	0.10	5.20	17.00	0.60	3.20	0.70	7.80	1.10	9.70
Tapioca	—	—	—	—	4.00	21.10	0.50	7.00	0.80	15.80	—	—
Onion	7.80	17.90	NA	NA	NA	NA	NA	NA	0.20	0.70	0.10	0.20
Ginger	NA	NA	0.70	1.20	7.40	45.30	2.60	20.40	0.50	4.00	1.00	1.90
Turmeric	10.10	7.00	NA	NA	1.40	6.40	0.40	3.60	NA	NA	1.40	2.80

* Based on *Basic Statistics of North-East India* (2000), North-Eastern Council, Shillong.

1. Low level of awareness (16%) and inadequate knowledge about its application, utility, method of operation and maintenance.
2. High initial cost, non-availability of basic finance.
3. Operating problems such as:
   - non-availability of spares/micro-tubes;
   - high cost of component;
   - leakage due to damage of pipes/drippers;
   - blockage/clogging of emitters/micro-tubes. The clogging could be due to algae, salt accumulation or foreign particles;
   - improper after sale services of commercial manufacturers;
   - lack of technical know-how, like the quantity of water to be applied for different crops under different situations, etc.
4. Threat of environmental pollution due to use of non-bio-degradable plastic material.
5. It is not suited for frost protection or for cooling during periods of hot weather.
6. They are not suitable for supplemented irrigation of large areas.
7. In windy dry regions micro-irrigation of young tree plants in light soils encourage soil erosion due to very limited wetted area. Strong winds also create problems of anchorage.

Design of Drip Irrigation System: Principles and Considerations

The following steps are followed for the efficient design of the drip irrigation system:

1. System capacity
2. Selection of emitting devices or drippers or tubing
3. Selection and design of laterals or tubes
4. Selection and design of sub mains
5. Selection and design of main lines
6. Selection and design of filtration system
7. Selection and design of pump unit

1. *System Capacity*

The system shall have the capacity adequate to fulfil 90 per cent of daily or peak water requirements within the designed area and in a stipulated operation period or not more than 16 hours of operation per day. The system capacity should include an allowance for losses of water application period. The capacity of the system should be matched with the available quantity of water in the source. In case of pond, canal, reservoir, system should be designed for incoming flow of water to outgoing flow with retention (dead) storage capacity.

2. *Selection of Emitting Devices or Drippers or Tubing*

There are numerous varieties of drippers/emitting tubes available, with different discharge rates, features, functions, characteristics and suitability to various crops. The selection of emitting devices are based on peak water requirement of crop, age and root zone, soil type, topography, soil water holding capacity and infiltration rate, hydraulic conductivity, life expectancy and cost economy. Dripper numbers and discharges are summarised in Table 9.5 for different crops and soil condition in order to achieve better irrigation and better yield of crops. However, an ideal or perfect emitting devices or drippers should meet the following requirements:

- Should be inexpensive, durable and serviceable (openable if possible).
- Should have relatively low discharge rate to keep the system cost low.
- Should have higher emission uniformity (90%) and low coefficient of variation (CV 5%) or as per Indian Standard.
- Should have relatively large cross-sectional area and flow path to avoid clogging.
- Should preferably have turbulent flow path and pressure compensation action (if possible).
- Should not create runoff within the immediate application area.

3. *Selection and Design of Laterals*

Laterals are the conduits which carry water from submain and feed to the emitters and are available in different sizes 12 mm, 16 mm, 20 mm, etc. Emitting devices/drippers can be fitted at determined spacing as per the requirements of crops. In case of Inline Drip Tubing, dripper intervals and discharge has to be determined before designing of system. As the laterals are significant with regard to pressure variation, proper evaluation of frictional head losses in the laterals is essential to achieve higher uniformity. Pressure variation 5 to 20 per cent and discharge variation 5 to 10 per cent throughout the system should be maintained in a range. The size, allowable length and frictional losses of laterals have to be determined by nomograms or charts and design guidelines provided by the manufacturer for their specific emitting devices.

4. *Selection and Design of Submains*

Submain is a conduit (PVC/HDPE) which carries water from main line and distributes among the laterals. The size, length and frictional head losses of submain have to be determined by nomograms or charts and design guidelines provided by the manufacturer.

5. *Selection and Design of Mainline*

Mainline is a conduit which carries water from source to the submain. While designing the mainline following points should be kept in mind:

- Permissible velocity : Should not exceed 1.5 metre per second.
- Frictional losses length of pipe : Should be limited to 5 to 10 metre per 1,000 m.
- Economic size : Should be such that, low initial investment, low annual cost and low power cost.
- Elevation & Class of pipe : Minimise use of higher pressure rating (class) of pipes at the elevated ground. Run the mainline as straight as possible in the shortest and at the same elevation.
- Control measures : Provide Air Release-*cum*-Vacuum Breaker Valves, Non Return/Check

Valves, Pressure Relief Valves, and Pressure Regulating and Sustaining Valves at appropriate locations.

6. *Selection and Design of Filtration Unit*

The Drip Irrigation Systems are characterised by large numbers of emitting devices, having fairly smaller flow paths. Because of the smaller flow path and orifice, they are more susceptible to clogging due to presence of physical impurities, organic/inorganic, silts, clay, sand, suspended particles, etc. a properly designed filtration system is the key to long-term successful operation.

Normally three types of filtration systems are used, viz., Hydrocyclone, sand filter and screen filter.

(a) *Hydrocyclone Filter:* It is also known as Centrifugal Filter or Sand Separator, and is used to separate the sand, silt or particles heavier than water but cannot remove algae, fibres, clay, etc. It is used at primary level where considerable concentration of silt, sand is present in water source and in case of canal, dam or reservoir.

(b) *Sand Media Filter:* Sand media filters are the most effective in removal of all types of physical impurities—organic/inorganic, algae, silt, clay and suspended particles, etc. Sand filters are recommended in case of open well, river, canal, dam water sources and where the quality of water in source is expected to vary with seasonal rainfall and runoff.

(c) *Screen/Mesh Filter:* It is normally used in case of clean water source with no presence of algae, clay and suspended particles.

7. *Selection and Design of Pump*

Pump unit must be capable of supplying required pressure and discharge for efficient functioning of system. Pump unit can be selected and designed by calculating total head and discharge required for efficient operating of system. Total head required for the system is calculated as:

(Suction + Delivery) + Filtration losses + Frictional losses in main line + Operating pressure + Fittings losses + Ventury head + Elevation.

Suction : Vertical distance between water level to centre of pump.

Delivery	:	Vertical distance between centre of pump to ground level.
Filtration losses	:	Frictional head losses in different types of filters. It is assumed 2 m for each type of filter unit, i.e., Hydrocyclone, sand and screen filters if three of them are connected in series then sum of frictional head losses of these three filters (6 m) must be accounted in calculation.
Main line losses	:	Frictional head losses occur in main line as per calculation.
Operating pressure	:	It is the recommended operating pressure given by the manufacturer at which system has to be operated and designed. In general 10 m (1 kg/cm^2) pressure is required for Non-Pressure Compensating Drippers.
Fittings losses	:	Frictional head losses occur in fittings like bends, elbows, tees, reduces, valves, other fittings, etc. and is assumed to 2 m overall.
Ventury head	:	Certain pressure is required to operate ventury or fertiliser applicator as recommended by the manufacturer. It is assumed to 5 m for manually operated ventury.
Elevation	:	Vertical distance between ground level near to source (as given above) to the highest level of ground.

With total head and discharge required, we can calculate horse power required by the pump for efficient operating of Drip Irrigation System.

$$H.P. = \frac{Q \times H}{75 \times a \times b} \qquad \text{...(i)}$$

where

Q = Discharge required in litre per second
H = Total head required in metre
A = Efficiency of motor (assumed 85%)
B = Efficiency of pump (assumed 80%)
$H.P.$ = Calculated horse power.

Table 9.5: Recommended Numbers and Discharge of Drippers/Micro Jets/Micro Sprinklers for Different Crops, Spacing and Soil Type

Crops	Spacing row & plant (Ft)	Dripper per plant × discharge (Lph)			* Micro jet per plant × pattern × discharge	Micro sprinkler	PWR**
		Loamy sand	Clay loam	Clay			
Grapes	6 × 6	2 × 4	2 × 4	2 × 4	1 × HC × LD	One for four plants	14
	8 × 8	2 × 4	2 × 4	2 × 4	1 × FC × LD	One for four plants	16
	10 × 10	2 × 4	2 × 4	2 × 4	1 × FC × LD	One for four plants	14
Pomegranate	12 × 12	3 × 4	3 × 4	3 × 4	2 × HC × LD	One for four plants	50
	15 × 15	4 × 4	4 × 4	4 × 4	2 × HC × LD	One for four plants	60
Guava	15 × 15	4 × 4	4 × 4	3 × 4	2 × HC × LD	One for four plants	50
	18 × 18	4 × 4	4 × 4	4 × 4	2 × HC × LD	One per plant	60
Apple	15 × 15	4 × 4	4 × 4	3 × 4	2 × HC × LD	One for four plants	40
Orange	18 × 18	4 × 8	4 × 8	4 × 4	2 × HC × LD	One per plant	60
Lemon	20 × 20	4 × 8	4 × 8	4 × 8	2 × FC × LD	One per plant	80
Mango	20 × 20	4 × 8	4 × 8	4 × 8	2 × FC × LD	One per plant	120
	25 × 25	4 × 8	4× 8	4 × 8	2 × FC × LD	Two per plant	130
Banana	5 × 5	1 × 4	1 × 4	1× 4	1 × FC × LD	One for four plants	16
	10 × 10	2 × 8	2 × 4	2 × 4	1 × FC × LD	One for four plants	20
Papaya	7 × 7	2 × 4	2 × 4	1 × 4	1 × HC × LD	One for four plants	20
Coconut	20 × 20	4 × 4	4 × 4	4 × 4	2 × FC × LD	One per plant	80
	25 × 25	4 × 8	4 × 8	4 × 8	2 × FC × LD	One per plant	100

* Pattern: HC—Half Circle, FC—Full Circle, LD—Low Discharge, HD—High Discharge.

** PWR—Peak Water Requirement in litre per day per plant.

REFERENCES

AFC, 1998. *Evaluation of Drip Irrigation System*, Agricultural Finance Corporation Ltd., Mumbai.

Ashwani Kumar, 1996. *Production and Prospects of Drip Irrigation System in India*, Department of Agriculture and Cooperation, New Delhi, p. 329.

INCID, 1994. *Drip Irrigation in India*, Indian National Committee on Irrigation and Drainage, Ministry of Water Resources, New Delhi.

Jha, M.K., Kamii, Y. and Chikamori, K., 2001. "Irrigation and Water Management: An Indian Perspective", Rural and Environmental Engineering, *The Japanese Society of Irrigation, Drainage and Reclamation Engineering*, Vol. 40, pp. 46-66.

Samual, J.C. and Singh, H.P., 1998. Current Trends in Micro Irrigation Development in Indian Horticulture, *Workshop on Micro Irrigation and Sprinkler Irrigation Systems*, Central Board of Irrigation and Power, New Delhi, pp. IX-1.

Shiklomanov, I.A., 1997. *Assessment of Water Resources and Water Availability in the World*, The World Meteorological Organization (WMO), Geneva, Switzerland, p. 88.

Singh, H.P. and Samual, J.C., 1999. "Hi-Tech Commercial Agriculture and Horticulture—Trends, Constraints and Prospects", *Financing Agriculture*, Vol. XXXI (3&4), Special Millennium Issue, Agricultural Finance Corporation Ltd., p. 75.

Yewalekar, Dilip, 2000. *Applied Design Principles and Considerations for Micro Irrigation System*, paper presented in "International Conference on Micro and Sprinkler Irrigation System", 8-10 February, 2000, Jalgaon (India).

10

SURFACE COVERED CULTIVATION FOR HIGHER PRODUCTION IN NORTH-EASTERN HILL REGION

— K.N. AGRAWAL, R.K. SINGH and K.K. SATAPATHY

Introduction

North-Eastern Hill Region of India covers a total geographical area of about 2.0 lakh sq km sharing of 5.9 per cent of country's total area. It has a forest cover of 26 per cent to the total geographical area and represent 50 per cent of the total flora in Indian sub-continent. The region is characterised by diverse physiography; ranging from hills and mountains to valleys of various dimensions. The annual average rainfall in most part of the region is more than 2000 mm. The agro-climatic conditions of North-Eastern region is representative of sub-tropical agriculture and horticulture under rain fed conditions. The overall agro-climatic conditions offers a good scope of growing a host of horticultural and agricultural crops. However, the excess rainfall takes away the fertile topsoil and nutrients, while high humidity and less sunshine period extends the cropping period, vigorous weed growth and higher incidence of insects and pests thereby reducing the yield. During winter season, the minimum temperature ranges between 0 to 6° C in most parts of the region, which adversely affects the growth of plants, resulting in farmer's reluctance to go for rabi crops or vegetable crops. The poor infrastructural facilities of transport, storage and marketing increases the cost of fruits and vegetables manifold and availability round the year is also not ensured. Food and nutritional security of the ever-increasing population is major concern with limited natural resources like land and water. There are two ways to increase the production; increase in either productivity or the cropping intensity. Since productivity of existing cultivation practice is very low so as to meet the food requirement, larger area needs to be brought under cultivation resulting in more damage to natural resources like land, forests, vegetation, etc. Apart from these, seasonality of the farm production increases the inter-State import and heavy fluctuation in market price. The loss and deterioration of the farm produce from the grower to consumer increases manifold. The quality and quantity of agricultural produce is closely related to environmental parameters and efficient utilisation of natural resources and inputs. Plastics offer a wider scope to grow horticultural crops under adverse climatic conditions. The experience of European countries to grow vegetable crops inside polyhouses or using plastic mulch to conserve moisture has become profitable enterprise. It is here that polyhouse technology plays a vital role. This controlled environmental condition may help in raising vegetables and horticultural crops in adverse climate, while improving crop productivity.

Green House Technology

The technology of green house has been popularised by various agencies for production of some selected crops having higher export potential. Initially the structure was made of glass. The initial cost of micro-climate controlled glass houses was very high. The plastic covered structures were developed initially as cheap substitute for glass house. The plastic green houses became popular around Mediterranean areas, where simple green houses helped in production of vegetables. Subsequently, the green houses were developed for tropical desert conditions and most recently for humid tropical regions. Although production cost of plastic is high, but often it proves economically viable as it enables to overcome the limitations of climate. Some standards have been developed regarding green house production technology through research backup carried out across the globe. For Mediterranean countries in the northern hemisphere the standards was summarised by Baille (1981):

- East-west orientation gives higher light transmission in winter and lower transmission in summer.
- Single span structure provides better light levels than multi span structure.
- An arched roof shape with higher radius of curvature gives improved light transmission.
- Protection from rain with a covering which should be as cheap as possible.
- The ventilation efficiency should be highest.

Design Considerations for Green Houses in NE India

Following considerations have to be given for adopting any design of green houses in NEH region of India:

- The green house should be low in cost and due to infrastructural bottlenecks; skills of working with structural material should be locally available.
- The material to be utilised for the structure should have enough strength and locally available. Considering the abundance of the bamboo and skill of the people while working with bamboo, it suits to be ideal material for structure for all categories of farmers.
- The cooling system should be able to function in high humid conditions. Since the fan pad cooling system will not work in high humidity conditions, natural ventilation is more viable and economical system to be adopted.

Designs Suitable for NE Region

Pipe Frame Poly House

This type of structure was designed by IARI, New Delhi with different dimensions of 4 × 24 m, 6 × 24 m and 8 × 24 m with 15-50 mm. The galvanised iron (GI) pipe was used as main structural material. The UV stabilised 150-200 micron film was utilised as cladding material. These structures are not suitable for the tall crops and suitable for the temperate conditions. However, due to GI pipe the cost remains slightly higher and needs more skill of welding and truss making.

Wooden Green House

The Assam Agricultural University (AAU), Jorhat has promoted three designs where bamboo is utilised as principal material for structure. The basic design is shown in Fig. 10.1.

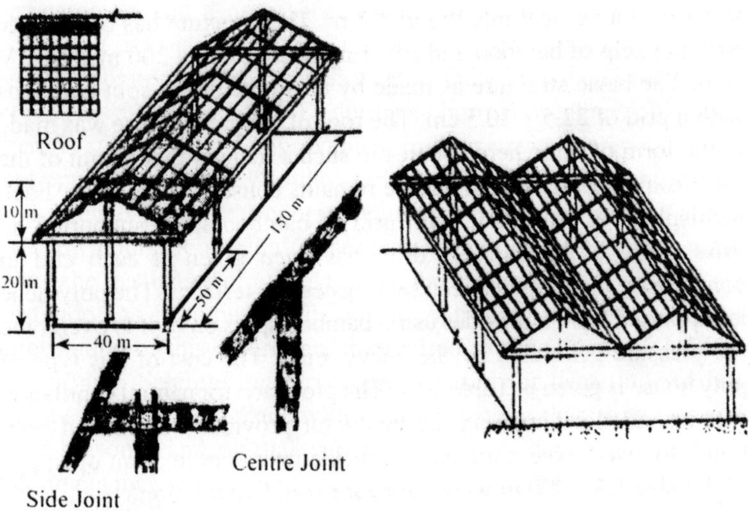

Fig. 10.1: Low Cost Green House-*cum*-Rain-shelter (after Barua, 2003)

(a) *Low cost green house*-cum-*rain shelter:* This type of structure serves as rain shelter during rainy season when vegetables cannot grow in open conditions due heavy rainfall.

(b) *Low cost double span green house*-cum-*rain shelter:* In case of more area to be covered under poly house, double/ multi span green house-*cum*-rain shelter can be used to reduce the cost of structural and cladding material. The design put forward by AAU has dimension of 18 × 11 m and oriented towards north-south direction.

(c) *Low cost roof ventilated poly house*-cum-*rain shelter:* In this type of structure the roof is made in two sections which meet at different level so that opening is created longitudinally along roof. The openings may be closed during winter to assist in raising temperature inside the poly house.

Low Cost Bamboo Poly House for Terraced Beds

A low cost poly house made of bamboos for bench terraced area has been promoted by the Division of Agricultural Engineering, ICAR Research Complex for NEH Region, Umiam. The total covered area of the constructed green house is 161.8 m^2 covering three bench terraces at a vertical interval of 1.2 m. The structure has been made with the help of bamboo and covering is done using 200 micron UV film. The basic structure as made by using the net of split bamboos with a grid of 22.5 × 30.5 cm. The roof of the green house was made in the form of three hemispheres in such a manner that height of the roof from ground of each terrace remains constant (Fig. 10.2). Each hemisphere is also made with grid of bamboo splits supported on truss made of bamboo. A drop has been given at each end to compensate the drop between two successive terraces. The polythene is supported from both sides using bamboo splits on roof to overcome the pressure generated by the heavy wind. The cost of this type of poly house is given in Table 10.1. The provision for natural ventilation was provided by opening the ventilator provided on sides of poly house to avoid excess increase in inside temperature. Two openings of the size 1.5 × 2.0 m were kept for natural ventilation.

Fig. 10.2: Low Cost Green House made for Terraced Land at Umiam

Table 10.1: Cost Estimate of the Low Cost Green House

Sl.No.	Particulars	Quantity	Total Cost (Rs.)
1.	Bamboo	125 nos.	2,500.00
2.	Labour Cost	LS	3,750.00
3.	200 micro UV Film	90 kg	13,860.00
4.	Rope, Steel Wire, Nails, etc.	LS	1,450.00
		Total	21,560.00
	Cost per m^2 of the floor area		133.25

Management of Green Houses in NE India

As high valued crops are being grown inside the green houses so management of soil, water and diseases, etc. are necessary to sustain crop production under intense agriculture in order to get higher return.

Soil Management

Soil is a major component for growth of any crop under the green house/poly houses. The soil *p*H, organic matter, fertility and moisture conservation are few major concerns to obtain the higher productivity levels. The higher temperature inside the green house reduces the *p*H of the soil. The fertilized soil shows higher reduction in soil *p*H because of proton produced by Ammonical fertilisers. Therefore, it is necessary to monitor soil after every three years in green house/ poly house conditions. The organic matter content is very necessary for growing various crops since it acts as store house of micro nutrients for plant growth. The higher temperature inside the green house leads to faster decomposition of the organic matter. Therefore, temperature control to a certain level is necessary for maintaining the organic matter content in the soil. The soil fertility deficiencies can be alleviated by the use of fertiliser. The mineralisation of the N takes place due to increase in temperature (Sharma and Agrawal, 1999). The soil fertility is best maintained if the temperature inside the green house remains within the 15-35° C, increase in temperature beyond this level leads to fixation of phosphates and potassium.

Disease Management

Due to extensive use of soil, which is not exposed to the natural

Table 10.2: Crops Suitable for Growth in the Green House/Poly House

Crops	Month of Sowing/ Planting	Variety	Suitable in States	Special Character
Tomato	For Spring Jan.-Feb. For Winter August	Arka Abha, BT 1, BT 10	Meghalaya, Assam, Manipur, Tripura and parts of Nagaland	Bacterial wilt and high temperature tolerant.
Capsicum	August-Nov.	Claifornia Wonder	All States	High yielder and somewhat resistant to bacterial wilt.
		Bharat		Susceptible to wilt.
Brinjal	June-July	RCMB-2, 3, 4, 6	All States	High yielder susceptible to stem and fruit borer.
Cabbage	Oct.-Nov.		Meghalaya, Manipur, Arunachal Pradesh, Mizoram	High yielder and suitable during winter season.
French Bean	July-Sept.	Bush type	Meghalaya, Arunachal Pradesh, Mizoram	The 45 days crop suitable for low temperature high humidity areas.
Cucumber	May for summer October for winter	Pusa Sanyog Local	Assam, Meghalaya, Manipur, Nagaland and Tripura	Best suited for crop on risers as creeper.
Bitter gourd/ Bottle gourd	June	Local	Assam, Meghalaya, Manipur, Nagaland and Tripura	Best suited for crop on risers as creeper.

biological factors, the soil born diseases, viz., bacterial wilt gets a strong foothold in the soil. Due to confined environment, non-pathogenic disorders also have been observed. The pathogens once introduced by any means build up their population very fast due to controlled environment and non-availability of bio-control agents. Some of the common diseases and their remedies based upon the experiences of many research workers who have worked in the region and at this institute as well as given below:

- *Bacterial wilt:* This disease is very common in tomato, capsicum and most troublesome in plastic green house conditions. The bacterial pathogens get ideal environmental for multiplication. Selection of variety is most important to withstand disease. The three important practice has been recommended by Gogoi *et al.* (2003): (i) application of lime as per soil test report to maintain *p*H of the soil, (ii) growing of Dhaincha on all sides of poly house and incorporating it in soil, and (iii) application of organic amendments such as oil cake and neem cake, etc.
- *Leaf Curl and Leaf Mould:* The leaf mould is a fungal disease while the leaf curl is caused by virus. The white fly is responsible for spread of this virus. The leaf curl can be cured only by indirect method of applying Carbofuran in soil @ 3g sq.m. at time of sowing and @ 1g/plant at the time of transplanting. For control of leaf mould the leaves should be clipped and burned. Spraying of Mancozeb @ 2.5g/lit gives better control (Gogoi *et al.*, 2003).
- *Mealy Bug:* These white mealy insects live in colonies and suck saps from leaf, stems and fruits resulting in twisting, curling, etc. The water stress in the plant makes them more vulnerable to such attack; so proper irrigation shall be provided. Their population can be checked by plucking and crushing the colonies.
- *White Fly:* White flies are very problematic in green houses. They are carrier of many viruses and pathogens. The regular irrigation also helps in checking their population. Spraying of contact and systematic insecticide such as malathion 0.1 per cent, phosphamidon 0.03 per cent, etc. will help (Srivastava, 1993).

- *Fruit Borer:* The fruit borers were found to be very active in crops like tomato, brinjal, etc. in the controlled conditions. The caterpillars feed on the tender foliage, leaves, stems and fruits. The stems and fruits are bored and internal sap is damaged resulting in rottening. The extent of damage may extend up to 42 per cent. The collection and destruction of affected fruits and plant parts control the attack in initial stage however, several insecticides such as endosulfan 0.07 per cent, carbaryl 0.05-1 per cent are also found effective (Sardana, 2001).

Apart from these, many other insects and diseases may occur in the green houses. The specific diseases and insects need to be controlled as per advise of the pathologist/entomologist.

Economic Comparison of Different Types of Green House

The cost of bamboo poly house is only Rs. 133.25 per m^2 as compared to cost of steel/aluminium framed green house with artificial control costing about Rs. 2000 per m^2 covered area. However, the life expectancy of bamboo structure is 3-5 years as compared to 10-12 years of life expectancy of conventional green house. Benefits available from the crop production along with maintenance, etc. are compared as shown in Table 10.3.

Table 10.3: Comparison of the Low Cost Green House and Steel/Aluminium Framed Fully Controlled Green House

Particulars	Bamboo Green House	Steel/aluminium framed green house (without automatic control)
Cost of Green House (Rs.)	21,560	76,000
Floor Area (m^2)	161.8	50
Cost of Green House (Rs./m^2 covered area)	133.25	1,520
Life Expectancy (Years)	3-5	10-12
Annual Benefit (Rs.)	4,338	9,566
Maintenance Cost (Rs.)	216	760
Operational Cost	Nil	1,200
Interest @ 9%	1,940	6,840
Total cost other than production cost	2,156	8,800
Net Profit (Rs.)	2,172	766
Net Profit per unit area (Rs./m^2)	13.43	15.32

Plastic Mulching in NEH Region

Plastic mulching is a practice of covering the soil surface around the plants to make conditions more conducive for plant growth through *in situ* moisture conservation and weed control. Use of dry leaves, straw, hay, stones, etc. as a mulching material has been prevalent for ages. However, introduction of the LDPE film as mulch improved moisture conservation, higher soil temperature and elimination of weed growth thereby increasing the crop yield.

Advantages of Plastic Mulching

- Earlier plant growth and faster maturity are the most obvious benefits since the soil temperature up to 2 inches depth increases up to 6-8° C under black plastic. The spring vegetables can be harvested one to two weeks earlier.
- Plastic mulch helps in preventing the direct contact of the edible product with soil and splashing water, applied fertiliser, etc. thereby improving the quality of produce.
- The loss of the nutrients from the soil is subjected to leaching in light soils prevalent in the region. A plastic covered bed prevents rainfall from percolating through the soil and reduces the micro-nutrient loss.
- Since black plastic mulch does not transmit the light it reduces the growth of weed.
- The evaporation loss of water is reduced considerably in mulched field.
- The surface saturation of soil is shielded due to covering and tends to have well aerated soil for proper root growth of plants.
- Properly managed crop under mulching gives better yield than non-mulched area.

Disadvantages

- Removal and disposal of mulch material is time consuming and costly affair. Plastic mulch material does not readily break down so it cannot be submerged in the soil.
- The purchasing of plastic and its equipment for laying them on plant beds increases the initial cost.

- Plastic mulch requires constant monitoring and intense management practices which increases the cost of production.

Mulch Film Thickness for Various Crops

- Mulch film with thickness of 15-25 micron is suitable for short duration crops such as vegetables, ornamental flowers, etc.
- Mulch film with thickness of 50 micron has a life expectancy of one year and can be used for medium duration crops like papaya, pineapple, tea, coffee, etc.
- Mulch film with 100 micron thickness is suitable for fruit crops of perennial in nature.

Methodology for Laying Mulch in Field

For Basins

- Prepare basins of 0.5 to 2.0 m radius after removing weeds and bringing soil to fine tilth.
- Apply recommended doses of fertilizer and manures.
- Cut the film to size of basins.
- Slit from middle of the film till the centre and make a circular cut to accommodate stem.
- Insert the film from the cut end (in the direction of wind) to position the tree trunk in the centre.
- Fix the margins of the film with soil.
- Make half moon shape cuts on film at corners.

For Row Crops

- Cut the film to the required width (based on the width of row/ridge) and lay along the row, making holes depending upon plant to plant distance and soil.
- Dibble seeds/transplant seedling in holes.
- Fertiliser and irrigation to be provided in furrows.

Benefits of Plastic Mulch on different Crops

The benefits of plastic mulch has been evaluated by many researchers on different crops. The mulching attributes to increase in yield, saving in water and weed control. The benefits in the terms of these attributes for various crops is being presented in Table 10.4.

Table 10.4: Benefits of Plastic Mulch for Different Crops

Crops	Yield increase (%)	Water saving (%)	Weed control (%)
A. Cereals			
Maize	1.6	13.4	NR
B. Pulses			
Soybean	12	81	NR
C. Oilseed			
Groundnut	49	26	30
Sunflower	10	28	74
D. Fruits and Vegetables			
Banana	10	53	90
Tomato (with drip)	56	78	95
(Without drip)	33	NR	NR
Grapes	54	78	—
Cauliflower	65	NR	96
Okra	38	60	NR
Cabbage	71	—	—
Pineapple	124	—	—
Potato	NR	37	95

NR—Not reported.

Problems in Protected Cultivation

Protected cultivation though very old for other parts of the world, it is in infancy in this region. There is a need to exploit the potentials of the technologies generated in this field so far. However, various constraints and problems are associated with it which are as follows:

1. Standardisation of the polyhouse/green house structures for various agro-climatic conditions needs to be worked out.
2. The initial cost of the cladding material, i.e., UV film is very high and becomes limiting factor for the vegetable/orchard growers.
3. Instruments for environmental control inside green house are not easily available and their initial cost is very high.
4. Lack of awareness among farming community with regard to its utility and potential.

5. Technological gaps with regard to cultivation practices for various crops under different protected conditions.

REFERENCES

Anonymous (2001). *Annual Report of AICRP on Plastics in Agriculture*, Palampur Centre, Ch. Sarwan Kumar Himachal Pradesh Krishi Vishvavidalaya, Palampur.

Anonymous (2003). *Annual Report of AICRP on Plastics in Agriculture*, Umiam Centre, ICAR Research Complex for NEH Region, Umiam, Meghalaya.

11

CROP DIVERSIFICATION PATTERN AND DISPARITIES IN AGRICULTURE IN NORTH-EAST INDIA

— UMESH CHANDRA SAHARIA

Raising the variety of crops in a crop season is the 'crop diversification' of an area. Crop diversification is an indicator of multiplication of agricultural activities with intense competition for space. Diversity in the structural forms of agriculture in North-East India such as cropping pattern, livestock or agricultural enterprise shows the possibility to raise the variety of crops and livestock determined by natural conditions. Impetus to crop diversifications is generally made by mechanisation in agriculture, improvement in irrigation system and use of fertilisers, pesticides, insecticides, etc. high yielding variety of seeds, etc. On the other hand, disparities in agriculture are engulfed by inter-personal inequalities, socio-economic class differences and inter-spatial discrimination, which create obstacles for agricultural development in the region.

This chapter aims to study the patterns of crop diversification in finding the nature of change in cropping pattern for the last decade.

As for the methodology concerned, the study is mainly based on State-wise secondary data from the Basic Statistics of North-Eastern Region, 1995. Investigation for crop diversification have been carried out by the statistical technique of Gibb's and Mortin Index (1962). The disparity in agriculture is investigated by finding cropping intensity and composite index of agricultural efficiency.

Introduction

The cropping pattern exhibits a variety of crops for arable land which involves intense competition among various activities to acquire more space. The competition marks the magnitude of diversification, which is directly proportionate one. It also denotes that greater the number of crop combinations the greater is the degree of diversification. The magnitude of diversification exerts the influence of socio-economic process of development and techno-organisational capacity. The diversification or specialisation in agriculture is determined by natural conditions. There is the cultural characteristic to strengthen the crop diversification and crop combination for a good and progressive farm management.

In the Indian society, agriculture practices are still prevailing traditionally in a caste structure. It is an identification of great inequality to cause political and economic instability in the country. According to Myrdal, 'the entire structure of inequality is bolstered by caste system, colour line, ethnic discrimination, nepotism and the general set of social and religious taboos. The vicious circle of cumulative causation is thus perpetuated. Present institutional arrangements impede the rise of social regional and institutional mobility and perpetuate the persistence of segmentation in social and economic life'. In other sense, many social scientists are of the opinion that disparities are needed to achieve development in the backward countries where monetary infrastructures are limited.

With this concept, the present problem has been studied for natural and social parameters like climate, soil and irrigation, fertilisers, variety of seeds, pesticides, etc. to give an impetus to crop diversification in North-East India. Here, six main foodgrain crops like rice, wheat, maize, small millets, gram, tar, and other pulses are sown in 3,747.0 thousand hectares of arable land with 5,068.0 thousand tonnes of production and 1,353 kg/hec. yield of land productivity. On the other hand, five major oil seeds like groundnuts, seasame, castor seeds, linseed rape seed and mustard seeds are sown in 337.5 thousand hectares with a production of 527 kg/acre yield. Again, thirteen of major crops are available with their respective area under cultivation, production and kg per hectare yield in the States of North-East India (Table A.11.1 in the Appendix). The whole North-East India comprises a total geographical area 25,505 thousand hectares.

Objectives

This chapter aims to find out the patterns of crop diversification and the nature of change in cropping pattern in North-East India. Secondly, it seeks to find out the nature and magnitude, reasons and policy measures for disparity in agriculture.

Analytical Procedure

(i) The investigation of crop diversification have been analysed through the Gibb's & Martin Index of crop diversification read as follows:

$$\text{Gibb's \& Martin Index} = 1 - \Sigma\, x^2/(\Sigma\, x)^2$$

where 'x' is the percentage of total cropped area occupied by each individual crop hectare wise.

Disparities in agriculture are analysed by three variables like—(a) Proportion of net area sown, (b) Cropping intensity, and (c) Composite index of agricultural efficiency.

(ii) (a) Crop Intensity (CI) is calculated by the expression

$$\text{CI} = \frac{\text{Gross cropped area}}{\text{Net area sown}} \times 100$$

(b) Composite Index of Agricultural efficiency by the expression

$$CI_x = \frac{\Sigma\, y_{ij}\, C_{ij}}{\Sigma\, C_{ij}}$$

where $y_{ij} = \dfrac{E_{ij}}{E_I}$ (E_{ij} = yield of the ith crop in the jth area, E_I = yield of the ith crop of the State and C_{ij} = proportion of the area under individual crop to the gross cropped area of the region.)

(c) The extent of disparity is to be identified with the Co-efficient of Variation

$$CV = \frac{\text{Standard Deviation}}{\text{Mean}} \times 100$$

Also, it is desirable to find out the causes of disparity by some environmental, technological, institutional, infrastructural and social factors that have been identified. The magnitude of disparity of this parameters has been computed by co-efficient of variation and their inter-relationships are done by the correlation.

$$r = \frac{\Sigma xy - (\Sigma x \,.\, \Sigma y)/N}{\sqrt{[\Sigma x^2 - (\Sigma x)^2 / N]\,[\Sigma y^2 - (\Sigma y)^2 / N]}}$$

where x and y are variables.

Crop Diversification Pattern

The cropping pattern in 1994-95 is somewhat uniform. In Table 11.1, it is observed that the crop diversification is not a favourable trend. The high category above 50 per cent of the crop diversification occupies a very small area with 2.56 per cent of total cropped area. There is only State Arunachal Pradesh falls in the category due to the fact that 95 per cent cultivable area is under *jhum* or shifting cultivation. The moderate category of crop diversification occupies only 6.87 per cent of cropped area and cover three States—Meghalaya, Mizoram and Nagaland. Remaining Manipur and Assam occupy about 80 per cent of total cropped area in the low category and Tripura falls in the very low category having more than 10 per cent area.

Table 11.1: Crop Diversification Index (Gibb's Martin Method)

Sl.No.	Range of the category	Category	Crop Diversification Index (Area under categories in %)		Variation
			1994-95	1995-96	
1.	Above 0.50	High	2.56	1.96	–0.60
2.	0.35–0.50	Medium	6.87	7.98	+11.11
3.	0.25–0.35	Low	80.12	70.75	–9.37
4.	Below 0.25	Very Low	10.45	19.31	+8.86

Note: Calculated by the author.

Table 11.1 provides category-wise explanation of changing pattern of crop diversification for the year 1994-95 based on Gibb's & Martin method. Here cropping pattern of North-East India is tending towards specialisation of rice cultivation in the *Kharif* season and wheat, maize and gram in the *Rabi* season. Rice and wheat are the two principal crops in the respective seasons.

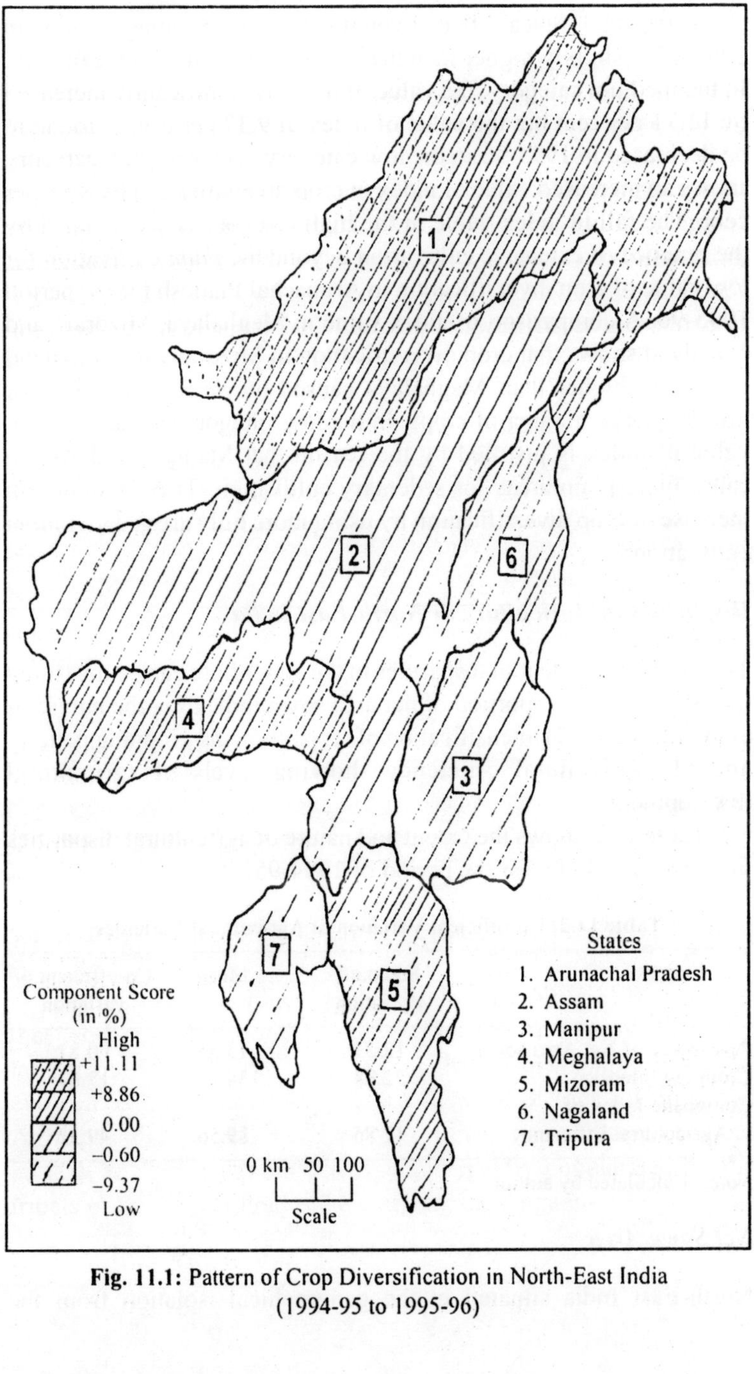

Fig. 11.1: Pattern of Crop Diversification in North-East India (1994-95 to 1995-96)

Table 11.1 reveals that diversification of cropping pattern in 1994-95 is slightly higher than that of 1995-96 in the high category. In the medium category, the value of index is significantly increased by 11.11 per cent but the value of index of 9.37 per cent is found to be decreased in 1995-96 in the low category. The very low category shows an increased value of index in crop diversification by 8.86 per cent. The fall in index value in the high category is accentuated by the practice of extensive cultivation beyond the *jhum* cultivation for low value of crop diversification in Arunachal Pradesh for the period 1995-96. A comparison of index value in Meghalaya, Mizoram and Nagaland shows that crop diversification in Mizoram and Nagaland is slightly higher than Meghalaya caused by *jhum* cultivation for growing large number of crops. In the low category, the decrease in value of index is justified by the reason that Manipur and Assam cause more plain areas for sedentary cultivation. Tripura shows an increase of crop diversification by using more hilly areas under *jhum* cultivation.

Disparities in Agriculture in North-East India

The nature and extent of disparities in the agriculture can be measured in respect of (1) proportion of net area sown indicating the nature of land utilisation, (2) intensification of crops showing cropping pattern, and (3) agricultural efficiency showing levels of agricultural development.

Table 11.2 shows the extent and nature of agricultural disparities in North-East India for the period of 1994-95.

Table 11.2: Co-efficient Variation of Agricultural Variables

Indices	Standard deviation	Mean	Co-efficient of variation
Percentage of Net Area sown	12.31	13.55	90.81
Cropping Intensity	22.98	134.87	17.04
Composite Index of Agricultural Efficiency	17.86	89.56	49.32

Note: Calculated by author.

Net Sown Area

North-East India situated in the geographical isolation from the rest of country has innumerable ridges, valleys, plateaus and eorded surfaces. Net sown area is predominantly identified by the various nature of land utilisation depending upon the physiographic features.

Jhum or shifting cultivation is practiced in the hill slopes of old plateau of Meghalaya and young folded mountains of Arunachal Pradesh, Nagaland, Manipur, Mizoram and Tripura while in the riverine plains of Assam, Manipur and Tripura settled agriculture is mainly practiced. Due to these facts, the disparity here is more acute in agriculture. The co-efficient of variation of proportion of net sown area, cropping intensity and agricultural efficiency score have been computed as high as 90.71 per cent, 17.04 per cent and 49.32 per cent. Therefore, agricultural development is not well balanced and well maintained in the economy of North-East India.

Crop Intensification

Cropping pattern is measured by the availability of technological inputs like irrigation, fertiliser, etc. But in North-East India, varied inputs prevail in the agricultural scenario and in the physical conditions like natural fertility, rainfall, floods, etc. The study reveals that the crop intensity in North-East India are three types—very high, high, and low.

Table 11.3: Proliferation of Crop Intensity

Types	Percentage of intensity	States
Very High	> 160	Arunachal Pradesh & Tripura
High	120-160	Assam, Meghalaya & Manipur
Low	< 120	Nagaland & Mizoram

The cropping pattern is determined by the availability of land which in Assam, Meghalaya, Manipur and Tripura is provided with alluvial soil with extra physical quality. But these areas do not have better irrigation facilities due to the frequent flood occurrences. On the other hand, low intensification of crops are found in, Mizoram, and Nagaland which is discernible in the hilly areas. Here, agriculture is mainly relied on rainfall and physical constraints. Arunachal Pradesh is found to be exceptional due to non-availability of correct information.

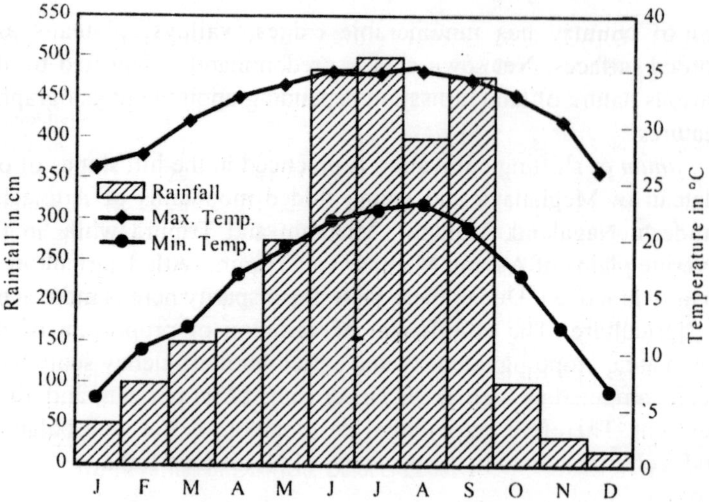

Fig. 11.2(a): Average monthly Rainfall and Temperature in North-East India (1986-95)

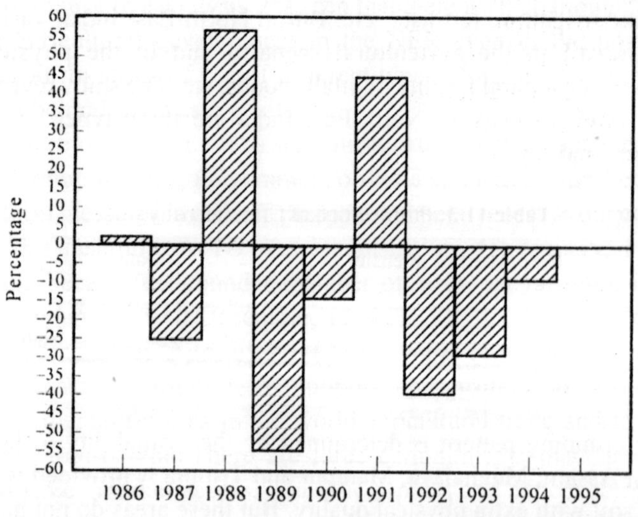

Fig. 11.2(b): Yearly Percentage Departure from Mean Rainfall (1986-95)

Note: Fig. 11.2(a) shows monsoon type of climate which is characterised by very high variation in rainfall and temperature during summer and winter seasons.

Fig. 11.2(b) indicates the trends in average annual rainfall from 1986-1995.

Source: Based on Meteorological Office, Borjhar Airport, Guwahati.

Agricultural Development

Assam, Meghalaya and Tripura have highest degree of agricultural development followed by Manipur and Nagaland where composite index of agricultural efficiency ranging between 100 to 20 per cent. These States have fertile alluvial soil in the plains and black and laterite soil in the hilly areas. The fertile areas of these States, with higher productivity of crops mainly cultivate single crop of area. Arunachal Pradesh, Nagaland, Mizoram and Manipur are not in the category where composite index of agricultural efficiency ranges 50 to 100 per cent. Areas in these States are affected by unfertile hilly soils and rainfall and have smaller proportion of net irrigated area. The inter-State variation of agricultural efficiency is minimum which indicates minor disparity in levels of agricultural development in North-East India.

Table 11.4: Correlation Matrix in the Agricultural Variables

Item	Net sown area	Cropping intensity	Composite index of agricultural efficiency
Net Sown Area	1.00	0.50	0.25
Cropping Intensity	0.50	1.00	0.25
Composite Index of Agricultural efficiency	0.25	0.25	1.00

Table 11.4 shows that the farmers of the States with large proposition of net sown area prefer to devote large area under the practice of multiple cropping. The land-use cropping pattern and agricultural development depend upon the nature of environmental, technological, institutional, infrastructural and social factors of the region (Table A.11.4 in the Appendix). The important fact is that there is not a single State where all favourable factors are prevailing largely or are very rare. Therefore, it is found that there is an unequal distribution of favourable factors and minimum inter-State variation in the proportion of net area sown, cropping intensity and composite index of agricultural efficiency in North-East India.

Conclusion

It is concluded that the environmental constraints and natural resources are localised in few States and technological inputs, infrastructure and facilities are very low and not equally distributed in specialised areas. But these unequal distribution do not throw significant influence on the concentration of agricultural efficiency as inter-State variation of composite index of agricultural efficiency is minimum in North-East India.

The computation of data which are made for 1994-95 give a specialised cropping pattern because of the high percentage of crop area that comes under low category of diversity index values. The change in cropping pattern could not be shown due to the non-availability of proper information. However, this type of crop diversification pattern has been studied to conclude that the cultural and physical factors are responsible for the change in North-East India.

REFERENCES

Agarwal, A.N. (1953). *Indian Agriculture and Its Problems*, Delhi, p. 45.

Ahmed Khan, Md. Zulfequar (1996). *Changing Pattern of Crop Diversification in West Bengal* (1977-78 to 1988-89), Delhi, pp. 171-74.

Das, Debashis (1996). *Disparities in Agriculture in West Bengal*, Delhi, pp. 145-46.

Jain, S.C. (1966). *Technological Changes and their Diffusion in Agriculture, Changing Indian Agriculture*, Mumbai, pp. 99-100.

APPENDIX

Table A.11.1: Distribution of Crops in North-Eastern Region

Items	Area under cultivation (in '000 hectares)	Production (in '000 metric tonnes/bales)	Yield (in kg per hectare)
Sugarcane	42.0	1,700.0	40,476
Cotton Lint	12.3	10.1	821
Jute	82.9	1,382.6	1,583
Mesta	14.4	75.9	949
Potato	96.5	688.9	27,139
Turmeric	11.5	12.0	1,043
Ginger	11.6	63.4	5,473
Soyabean	8.2	11.3	1,379
Onion	6.5	13.3	2,046
Chillies	25.0	17.7	708
Sweat potatoes	16.5	65.9	3,994
Banana	56.2	680.9	12,116
Other fruit crops	101.9	720.9	Not available

Source: Basic Statistics of North-Eastern Council of India, 1995-96.

Table A.11.2: Variation of Agricultural Factors in North-East India, 1995

States	Geo-graphical area (in hectares)	Net average (in sown thousand hectares)	Percentage of net area	Total	Cropping intensity (in %)	C.I. of agricul-tural efficiency
Arunachal Pradesh	8,374	149	1.77	247	165.77	50.50
Assam	7,844	2,706	34.49	3,797	140.31	120.23
Manipur	2,233	140	6.26	180	128.59	98.02
Meghalaya	2,243	202	9.01	243	120.29	110.56
Mizoram	2,108	65	3.08	74	113.84	70.25
Nagaland	1,658	190	14.54	210	110.52	59.26
Tripura	1,049	270	25.73	445	164.81	125.21

Note: Calculated by the author.

Table A.11.3: Co-efficient of Variation of Agricultural Variables

Variables	Standard deviation	Mean	Co-efficient of variation
% of Net Sown Area	12.31	15.55	14.81
Cropping Intensity	22.98	134.87	17.04
CI of Agricultural efficiency	17.86	90.43	16.92

Note: Calculated by the author.

Table A.11.4: Co-efficient of Variation and Correlation of Agricultural Variables in North-East India

Sl. No.	X Variable Environmental Factor	Co-efficient of Variation	Y Variables		
			% of net area sown	Cropping intensity	Agricultural efficiency score (*r*)
1.	Relative Relief (in metre)	259.61	–0.21	–0.61	–0.52
2.	Annual Rainfall (in mm)	100.02	–0.19	0.31	–0.75
3.	% of Alluvial Soil	42.31	0.18	0.59	–0.36
4.	Fertiliser Consumption	39.23	0.17	0.26	–0.46
5.	Percentage of irrigated area	120.13	0.21	–3.29	–0.62
6.	Percentage of area under high yielding variety seeds as Institutional Factor	65.35	0.40	0.03	–0.69
7.	Man Land Ratio	66.01	0.21	0.89	–0.31
8.	Percentage of literacy	25.92	–0.42	– 0.56	+0.59
9.	Percentage of net area sown	90.81			
10.	Cropping intensity	17.04			
11.	Agricultural efficiency score	49.32			

Note: All the above factors are computed by standardised value

$$= \frac{x - \bar{x}}{\text{S.D.}}$$

where x = variables, $\bar{x}$ = mean and S.D. = Standard deviation.

12

WOMEN AND AGRICULTURE
Their Overall Role

— SUSMITA DAS

Women are the invisible and unrecognised backbone of agriculture in any South Asian countries including India. The significance of their role can be measured not only by their high participation in farm and non-farm activities in rural areas, but also by their intimate connection to rural customs, tradition and values. In the fast changing population composition of rural areas, it is women, who keep the rural way of life still alive and also at the same time suffer from it. Their role varies from being an agricultural labourer, *de jure* landowner, *de facto* household head or a manager of their homesteads.

In spite of their round the clock activities, when it comes to their participation into economy, their work often gets unrecognised and thus goes unrecorded. This happens particularly in countries of South East Asia including India, where a host of factors independently or in combination exert their influences in officially keeping their name outside the record book. Some of the reasons responsible for such a dismal rate of female workers include the flaws in the design of questions and definitions. Besides in many of these societies the status of the family is measured *vis-a-vis* women's absence from gainful economic employment. In some countries (Pakistan) full time agricultural workers includes those only who perform agricultural work exclusively. The sex of the enumerator and the respondent also determines women's inclusion or exclusion from the list. All these often classify women as supplementing the work carried out by men, which is defined as 'actual' work; task performed by women exclusively as 'non-economic activities'; unpaid labour in terms of cash, as the only means of visible economic accounting. Thus women are considered and treated as the subordinate half of the family within the household and, therefore, only 'family helpers' and not as economic contributors.

In view of the above, an effort is made here to understand the role of women as a player in the economy. In view of the low participation of women in economic activities accentuated by their low status or otherwise which they have in many parts of the country, it is felt necessary to see their condition in these respects in a state of the country where women are said to possess a better status (if not higher) as the State is largely dominated by tribal population and, is largely egalitarian where there is an absence of social taboo against work. Moreover, being a matrilineal society, women's participation into work is not looked down upon. All these have made the selection of Meghalaya for the present study an ideal one. Again being an agrarian society with huge population into agriculture directly or indirectly, the emphasis of women's participation into economy and particularly their role into agricultural sector gets justified.

To understand women's participation into agriculture, a few selected variables have been considered which includes the proportion of females in gainful employment, proportion of females in agriculture both as cultivators and agricultural labourer. An effort has also been made to see the male-female differentials. The study is based on secondary data collected primarily from the 2001 census data conducted nationwide including in the State of Meghalaya.

Table 12.1: Percentage Share of Female-Male Workers in Meghalaya (2001)

District	Total women employed	Women as agricultural workers	Total man employed	Man as agricultural workers
Jaintia Hills	37.5	78.1	48.0	74.5
East Khasi Hills	27.8	36.3	45.2	29.0
West Khasi Hills	43.1	89.4	46.0	81.3
Ri-Bhoi Hills	40.2	81.8	54.1	69.8
East Garo Hills	40.6	85.6	46.4	80.0
West Garo Hills	32.1	81.4	49.5	66.6
South Garo Hills	43.2	83.3	51.6	67.3
Meghalaya	35.0	72.6	47.8	61.1

Female as Workers

Meghalaya, a matrilineal egalitarian tribal State with an absence of the existence of any taboo as far as female participation in economic sphere is concerned was expected to show a high proportion of female workers, but unfortunately it is not to be. In a country where less than twenty-six per cent of the female is categorised as workers, the State's position is found to be little better, though nothing extraordinary. This is surprising considering a comparatively high female literacy rate (60.4%) that the State enjoys and a general perception that almost everyone works, the share of females registered as workers works out to be only about thirty-five per cent of the total female population. Again, a huge percentage of them is engaged in the agricultural sector, meaning that the high literacy rate has little bearing as far as its influence in determining female economic participation is concerned. The two comparatively more urbanised district of East Khasi Hills and West Garo Hills have female workers' share which is below the State average while among the rest of the districts South Garo Hills have the maximum with a share of 43.2 per cent followed by East Garo Hills and Ri-Bhoi district. It, therefore, appears that higher the potential of employment in the agricultural sector the higher is the rate of female participation and *vice versa*.

Again if proportion of women as agricultural workers to total women employed is taken into consideration (Table 12.1), it is found then that barring the East Khasi Hills district, in all the other districts more than 80 per cent of the female workers are into agriculture.

Considering women agricultural workers as a percentage of total employed women against male agricultural workers to total employed male, the share of former exceeds from that of the latter, indicating that female participation in agriculture is far more than that of male. At the same time it also reflects a single dimensional work for female as against multidimensional nature of work which men perform. This further reflects a continuation of women's greater participation into agriculture and a restricted entry of them in other sphere of work particularly into the tertiary sector.

The percentage of male-female employed into agriculture at the district level as shown in Fig. 12.1 highlights the spatial variation that exists among the districts of the State. The gap between the sexes is far more revealing in the western part of the State where agriculture is the predominant occupation of the population as these areas have low level planes suitable for cultivation of a variety of crops including rice, which always need and employ more female hands. The same fact stands true for Ri-Bhoi district where too similar conditions explain the gap between the sexes.

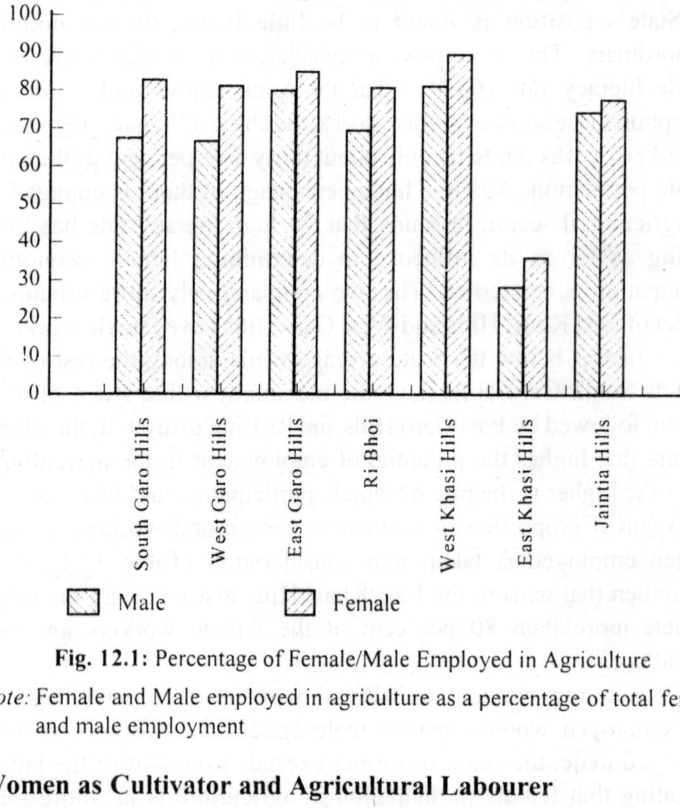

Fig. 12.1: Percentage of Female/Male Employed in Agriculture

Note: Female and Male employed in agriculture as a percentage of total female and male employment

Women as Cultivator and Agricultural Labourer

The involvement of women in agriculture is spread over a large number of activities. In fact in certain specific region they perform more task than men. Though the participation of female in this sector is determined by several factors including the type of activity, crop cultivated, nature of geographical area, socio-economic status, subsistence or cash crop, etc. there is no denying of the fact that their role in agriculture and agro-related activities is increasing day-by-day.

One of the leading reasons for their increasing participation is directly related to growing male selective migration that is happening from rural areas towards towns and cities of various sizes, leaving the women to come forward and replace the shortages of labourer in the field. This has resulted in growing participation of female in the agricultural sector both as cultivator and as agricultural labourer. Not surprisingly, therefore, the share of working female into agriculture is more than that of male's share into the same sector almost everywhere in the State. The gap though, wider in certain districts where rice is cultivated extensively, i.e., in the two Garo hills districts and in Ri-Bhoi district, the gap narrows down in other districts where a combination of crop is cultivated. The noticeable feature however, is that in all the districts of the State the gap is a significant one.

Table 12.2: Share of Women as Cultivators and Agricultural Labourers in Meghalaya (2001)

District	Cultivators	Agriculture Labourers
Jaintia Hills	43.4	34.7
East Khasi Hills	22.0	14.3
West Khasi Hills	65.2	24.2
Ri-Bhoi	60.9	20.9
East Garo Hills	70.1	15.5
West Garo Hills	61.8	19.6
South Garo Hills	66.8	16.5
Meghalaya	51.9	20.7

Now coming to the nature of work that women perform within the agricultural sector, it is found that a majority of them are engaged as cultivators and a significant number is employed as agricultural labourers. Although the share of women in the latter category is not high all throughout the State, it is the highest in the Jaintia Hills district. A cursory look at Table 12.2 showing the share of female as cultivator and as agricultural labourer reflect that districts with higher proportion of female as cultivator also have an equally high proportion of female engaged as agricultural labourer, indicating the entire female working population's engagement into agriculture alone. The fact stands true for all the districts of the State with the only exception being the East Khasi hills district where the overall share of female as workers is on the lower side and among them most are engaged in the service sector which is perhaps the largest employment generating sector. It thus indicates that in the State of Meghalaya, high female work participation means their engagement in the agricultural sector alone. Also it indicates that women are kept outside the other forms of employment available otherwise only to male members, reflecting the presence of some kind of sex selective nature of work.

Conclusion

The analysis above puts forth a number of questions with regard to women's work participation. Some of these questions have been raised frequently, though very little efforts have ever been made to answer them.

To begin with why is it that in the State where the literacy rate of female is high compared to most other parts of the country, it does not get reflected in the participation rate and nature of work that female performs. Does it then suggest that high literacy rate has little bearing when it comes to women's employment in the formal sector is concerned. Also, it may be the case that high literacy rate may not necessarily mean a high level of educational attainment and, therefore, an overwhelming number of them gets engaged in the agricultural sector alone. If that is the case, then, time has come to make a sincere effort to improve the level of educational attainment among the female section of the population, which will eventually resolve their unidirectional nature of work.

Again in the State where the involvement and contribution of female in agriculture in noteworthy, a relevant question that comes to mind relate to their access to land they cultivate. As land remains primarily with community, the question is whose land these female tend to cultivate and an even more pertinent question relates to the income that comes by selling the surplus products. Who enjoys the profit? If they happen to cultivate the land belonging to the community then are they given the right to take the income and the profit, or are they simply denied of it? In case of the latter, who takes the share of profit? Is it the male member of their family or somebody else? Is it like most other parts of the country where female work as unpaid worker? Their work is well utilised, but remuneration even when paid is the lowest. A far more serious problem that arises out of this situation is that in the absence or lack of ownership these women are denied of having any access to a number of agricultural support services.

Another aspect, which demands an answer relates to women who work as cultivator and agricultural labourer, is to know whether these women have an access to market outside their villages or not. As most of these villages have market on a weekly basis, do they then sell in these markets only or do they themselves transport the surplus product to the nearby town/city market on their own? This also reflects their overall role as a player in the economy as women who do not sell their product themselves usually do not have control of the income generated from these sales.

All these questions when answered can only bring out the role that they play in agriculture and, therefore, their overall role in the economy. The State which boosts itself to be an egalitarian society can only hope to accord a greater acceptance of women as true workers both in terms of a genuine worker and an earner to allow them to play a crucial role not only to improve their own living but also the lives of the entire household. A better economic status is enough to ensure a better and healthier family, as women's income contribute more to household food security than when the income lies in the hands of men. Moreover, in the State where female generally is found to remain more responsible towards their children, better financial position will assure a better future for children and, therefore, a better scenario.

REFERENCES

Das, Susmita, *Rural Poverty and Poverty Alleviation Programmes: A Case Study of Lynden and Urksew-Wahpathaw Villages*, unpublished M.A. IVth semester project report, NEHU, Shillong, 1993.

Human Development in South East Asia 2002, *Agriculture and Rural Development*, published by Mahbub ul Haq Human Development Centre, Islamabad, Pakistan, 2003.

Okafor, F.C., "Population Pressure, Agricultural Change and Environmental Consequences in South Eastern Nigeria", in *Geography of Environment*, ed. Md. Shafi, *et al.*, Rawat Publication, Jaipur, 1994.

13

AGRICULTURAL PLANNING AND INFORMATION BANK (APIB)
A Social Enterprise for the Unemployed in NER

— P.P. NAGESWARA RAO

Introduction

The advancements taking place in biotechnology, including cell and tissue culture, and embryo implantation are providing better seeds, drought resistant and high yielding strains of crop as well as cattle. Newer methods of crop rotation, irrigation, better use of fertilisers and right agricultural practices are now available for sustainable agriculture. Farmers have adopted these technologies and produced the best possible. However, there have been instances of severe crop losses due to drought, crop pests and diseases, land degradation; environmental change, etc. Farmers are facing a lot of hardship due to uncertain market and weather. They produce for unknown markets and do not have access to accurate weather information to plan their activities. Crop and cattle production strategies have to be more innovative than today to get the best per unit of land and water used. Agricultural planning (selection of crops, cropping patterns, input mobilisation, etc.) has to be done based on long-term rainfall, temperature, soil factors such as texture, structure, depth and topography. So far, the producers are protected by giving them remunerative support price and the consumers are supplied the essential commodities through the public distribution system. As a result, the agriculture sector is able to sustain the economic growth. This support is slowly waning. In the long run, the growth of foodgrains and agriculture economy will depend on the level of technology and knowledge in hand and its adoption by several farm families, the decision-makers at the grass-root level.

It is in this context, precise scientific information on characteristics, potential, limitations and management needs of land, water resources and climate is indispensable. Here we discuss the applications of remote sensing (RS), Geographical Information Systems (GIS) and computer aided information processing and dissemination tools in increasing the productivity of agriculture and allied sectors.

Planning Advice and Information Support

Remote Sensing techniques could be used (i) to estimate the production of major crops one month before they are harvested and for demarcating areas requiring fertilisers, irrigation, crop protection and crop rotation, (ii) to monitor the time of harvest of a crop so as to break the life cycle of the pest, (iii) to estimate the acreage under alternate hosts for assessing the carry over potential of the pest in the off season, and (iv) to identify the areas affected by pest/disease. In other words, the spatial and temporal variations in crop yields could be quantified and precision agriculture could be practiced. The GIS tools could be used for (i) integrating the natural resource supply with the demand, (ii) assessing the adequacy of the infrastructure for storage, processing, marketing, etc. (iii) trend analysis and gap area identification, etc.

Among the highly dynamic information needs that the farmers most often seek are weather and market. Both these variables are beyond the grasp of the farming community. They produce for unknown markets under unpredictable weather conditions. Using the remotely sensed data from Indian Remote Sensing Satellites (IRS) having capability of resolving what is grown in one-tenth of an acre, it is possible to estimate the area under foodgrain crops, horticulture and commercial crops of neighbouring States and countries. Thus, a sort of agriculture intelligence and market information can be supplied to the farmers. Similarly, the remote sensing payloads from geostationary satellites like Kalpana-I, INSAT, etc. can provide hourly information on weather systems that are approaching a place and accordingly farm operations can be advised.

An innovative approach of integrating remote sensing, GIS and IT in providing services to the farming community has been demonstrated through a project called Agro-Climatic Planning and Information Bank (APIB). It is a single window access to agriculture knowledge and a synthesiser of spatially referenced agriculture data. It uses RS for assessing natural resources endowments, GIS for addressing adequacy of agriculture infrastructure, Relational Data Base Management Systems (RDBMS) for supplying farm knowledge (package of good practices, information on high yielding varieties, etc.), also for knowing the trends in prices, rainfall, crop productivity, etc. (Fig. 13.1).

Fig. 13.1: A Schematic of APIB

The first step in developing APIB is to conduct a district-wise primary survey on the information needs of the farming community; identify the needs that are most important and essential; categories those that are highly dynamic and required to be supplied during various stages of crop growing season and that are static one time need. The second step is to take stock the potential of the natural resources of the district, block or a group of villages using remote sensing techniques and evaluate their availability, accessibility and usability for meeting the food, fodder, fuelwood and fibre needs of the human and bovine population of the area under study. Then, using the tools of GIS, the maps of natural resources (the supply components) and socio-economic parameters (demand components) are integrated to delineate coherent land units that can be allocated for appropriate land-use, cropping systems and farming systems. The third step in developing APIB is to convince the people at the village level on the benefits of alternatives to the currently practiced land-use pattern. The satellite based instructional television and radio can play important role in carrying this message. APIB provides planning advise and information support in local language.

This bank can be built with the substantial information already generated district-wise through various projects. The bank also relies upon the extensive organisational linkages established in the collection of data from the universities and research institutions specialised in agricultural sciences, State Departments of Agriculture, and Department of Horticulture, State Agro Industries Corporation, State Agricultural Marketing Board, State Directorate of Economics and Statistics, various non-government and private organisations in the State and outside could be pooled and synthesised into planning advises using GIS and RDBMS tools.

Planning advisories and support information that can be retrieved from APIB include: crops that are appropriate for a farmer's land and water resources, variety/hybrid selection for principal crops, comprehensive package of practices from public and private sources, fertilisers, plant protection, cost of cultivation and market prices, area under cultivation and yield, dairying and insurance, information on time of arrival of sowing rains, crop growing periods and weekly probability of rainfall.

Agricultural Planning based on Rainfall Analysis

Analysis of the water budget of a given area based on the amount of precipitation (P), evaporation (E) and run-off (R) allows delineation of a place into climatic regions. Such an approach allows us to assess the conditions favourable or not to the growth of plants (natural and cultivated crops). The agro-climatic zoning serves as a general guideline on the choice of farming systems to be adopted by the farmers.

The length of Crop Growing Period (CPG) of a place can be estimated by plotting precipitation (water supply) and evapotranspiration (water demand) on a weekly or monthly calendar and the period (in days or weeks) during which at least 50 per cent of the demand is met by the supply is identified. Yet another important criterion in deciding the suitability of a place for a particular agricultural practice is the degree of year-to-year reliability of rainfall, or rainfall variability. Long-term trends in rainfall and temperature can be used to appropriately design the crop selection to avoid the water shortages during critical crop growth periods. The farmers of the region could use such analysis in deciding upon the choice of crops and cropping systems.

Another useful information for the farmers is the mean dates of commencement of sowing rains (CSR). The CSR is a short spell of rainfall at least 25 mm per week with 0.6 mm or more in any five days of a week. It was found by IMD's Agromet Division that the sowing rains occur much ahead of the normal onset of monsoon in the North-Eastern Region due to occurrence of pre-monsoon thunderstorms. The CSR commences from mid April and that is the beginning of agricultural season that coincides more or less with the thundershowers of April. Farmers could make use of this information in the choice of crop varieties to be planted in a year.

Traditionally farmers refer to the lunar calendar in which the twenty-seven constellations through which the Sun passes in a year and they expect the likely period and quantum of rainfall. The period of each epoch is about 14 days. The knowledge of an experienced farmer about rainfall during these epochs could help him in mobilising his resources during each farm operation. There are many proverbs and folk songs linking weather and agriculture. One can adopt this farmer-friendly language and analyse the probability of occurrence of rainfall for each epoch. This crop-weather window will not have any stress on crops. Such information can be synthesised for each epoch and used in farm planning. By superimposing satellite-based rainfall assessment and forecasts (now available freely on the Internet) over these crop-weather calendar farmers can plan their resource mobilisation and input management operations.

Farm Operations based on Weather Information

So far we have discussed the agro-climatic information farmers can use in their long-term and seasonal planning. That is not sufficient. The farmers also need information on weather conditions during various stages of crop growth. The farm operations could be advanced or postponed and if necessary hastened if weather ahead of a few hours and days is available to the farmer. Although such forecasts are incorporated into the farmer's weather bulletins issued by Regional Forecasting Offices of IMD during abnormal weather situations, there is still scope for improvement in their area coverage, content and applicability to day-to-day agricultural operations. Application of pesticides in the IPM context requires weather information ahead of a few hours. Rainfall after pesticides spray or dusting may wash away the pesticides or at least dilute the recommended dose of the pesticide and becomes inefficient in controlling the pest. Weather information of this nature is also required to schedule crop harvest and drying. Satellite imagery could be used for detecting various cloud signatures associated with weather systems that bring rain and their seasonal and diurnal variation. Farmers having some basic training in meteorology and understanding of interpretation of clouds from satellite imagery can infer the impending weather and carry their farm operations.

Information Dissemination as Livelihood

All the information and planning tools thus derived can be computerised and the APIB can be run as an enterprise starting with an investment of about Rs. 50,000, a large portion of it through bank loans or subsidy. APIB can be built around a personal computer. Information and planning advice can be sold as a commodity. A quick estimate of employment opportunities shows that nearly 5,00,000 graduates per year can get livelihood on running an APIB like information kiosk.

The planning advisories and information services thus generated can be reached to the farmers and other decision-makers by post as well as through the Community Information Centres (CICs), the regional television and radio stations, the print media especially the journals and magazines on farming and related fields. The Training and Development Communication Channel (TDCC), a simple one way video and two way audio teleconferencing network supported by Indian National Satellite (INSAT), could be another means to reach the large farming community. These facilities are available at the North-Eastern Space Applications Centre (NESAC), at Umiam (Barapani), 18 km from Shillong, off G.S. Road.

14

AGRICULTURE IN ASSAM
Emerging Micro-trends

— GORKY CHAKRABORTY

INTRODUCTION

The WTO regime is yet to resolve the agrarian question. While the developed and developing countries are trying to settle their scores over various agricultural issues during several ongoing rounds of trade negotiations,[1] agriculture itself (particularly in the developing nations) is witnessing far-reaching changes. The 'pulls' and 'pressures' of neo-liberal globalisation is re-defining the contours of the erstwhile inward-looking and self-sustaining(!) rural societies. This is manifested *via* crop diversification, commercialisation and corporatisation of agriculture. Various aspects concerning crop selection, cropping pattern, preference for non-traditional export oriented crops as well as issues related with food security are being discussed at several quarters. Opinions are diverse and so is the process of diversification itself.

Changes in the rural economies can be brought about through number of related factors.[2] It might occur within the agricultural sector where an alternative crop production (often referred as cash crops) makes a difference. It may also be related to livestock rearing within the broader framework of the agricultural sector. Secondly, changes can be brought about through a combination of agricultural and non-agricultural activities, i.e., fostering growth in both farm and non-farm sectors within the rural economy. Again, by means of extensive linkage between the rural and urban regions, e.g., through transfer of rural population to urban areas for productive employment and/or through their remittances from urban areas, there can be changes in the rural areas.

In this chapter we limit ourselves to the first aspect mentioned above, i.e., changes in rural areas through a change in cropping pattern. Attempts have been made to understand the process of changes in rural Assam which has been brought about by cultivation of different crops. Both crop selection and response of the producers have been diverse in various regions of the State which is analysed in the following sections.

CHANGES IN AGRICULTURE

Changes within the rural economy can take various forms, changes in cropping pattern is one among them. But even this change in cropping pattern is far from uniform in nature. Assam presents us with a varied picture. While in Upper Assam change in cropping pattern has essentially meant the growth of small tea gardens, in certain parts of central Assam the shift has been towards floriculture, while in the remote riverine tracts of the river Brahmaputra (locally called as *chars*) the shift has been towards cultivation of new variety of paddy. The reasons for this diversified situation are analysed here. While in case of tea, macro-level secondary data is used, in case of floriculture and paddy, micro-level primary data are used as our source of analysis.

Small Tea Gardens in Upper Assam[3]

Assam is synonymous with tea both at domestic and global level. Tea as an economic venture has flourished in Assam since 1840 and has been the most important aspect of revenue generation till date. However, tea has always been a richman's venture both during the colonial period and the post-independent period.[4] It was only during the post-emergency Janata government that the importance of small tea gardens in revitalising the economy of the State was highlighted.[5]

During this period few paddy cultivators of Golaghat, Jorhat, Sibsagar and Dibrugarh districts initiated the process of tea cultivation in their small holdings, ranging between 0.13 hectare to 3 hectares. But with the fall of the Janata government, the scheme failed too. It was only after the formation of All Assam Small Tea Growers'

Association during 1987 and simultaneous research and consultancy by Department of Tea Husbandry and Technology (now called as Small Tea Growers Advisory Programme) of the Assam Agricultural University, Jorhat, small tea gardens[6] caught the fancy of enterprising section of the local population in this region.

Number of Small Tea Gardens

During 1987 when the Grower's Association was formed there were only 21 small tea growers in Assam but thereafter the growth has been phenomenal. During next three years it shot upto 657 and upto September 1999 there were 19,494 small tea gardens spread over 20 districts of Assam. Dibrugarh district has the highest number of growers followed by Jorhat, Sibsagar, Tinsukia and Golaghat. Over 90 per cent of the gardens are located in these 5 districts (See Table 14.1).

Size of Holdings

Majority of the holdings (34.68%) are in the 10 to 19 *bigha* category. About 25.63 per cent are in 5 to 9 *bigha* category and 14.17 per cent are in 20 to 29 *bigha* category. Nearly 85 per cent of all the holdings are below 30 *bigha*. Only 4.6 per cent of the holdings are above 50 *bigha* (See Table 14.2).

Area

It is estimated that 42,774 ha of land is under small tea cultivation. Dibrugarh district has the largest number of small tea garden both in terms of numbers and area involved. About 27.68 per cent of total land under small tea garden is in Dibrugarh. Golaghat occupies fifth position in terms of the number of gardens but it occupies second position (17.16%) in terms of area involved. Tinsukia, Sibsagar and Jorhat districts follow accordingly (See Table 14.3).

Land-Use Before Plantation

Small tea gardens thrive on various sorts of land including own land, ceiling surplus land, grazing land, government land, reserved and unreserved forest land, etc. Encroachment is common in all varieties of land except for own *patta* land.

Table 14.1: Estimated Growth of Small Tea Growers in Assam (1990-91 to 1999)

District	1990-91	1991-92	1992-93	1993-94	1994-95	1995-96	1996-97	1997-98	1998-99	Sept. 1999
Tinsukia	12	159	387	528	609	738	951	1,342	2,153	2,595
Dibrugarh	24	312	915	1,635	1,998	2,499	2,940	3,846	5,856	6,867
Sibsagar	216	273	576	1,005	1,236	1,320	1,551	2,049	2,232	2,760
Jorhat	156	297	486	639	804	975	1,401	1,869	2,502	3,174
Golaghat	237	477	726	837	1,221	1,383	1,536	1,698	1,881	2,148
Karbi Anglong	3	6	51	63	66	75	90	96	102	102
Nagaon	3	3	6	9	18	39	48	63	81	93
Marigaon	—	—	—	6	7	16	18	20	20	22
Kamrup	—	1	1	1	3	18	18	71	77	77
Nalbari	—	—	—	—	—	—	—	—	—	10
Goalpara	—	—	—	—	216	216	216	219	222	246
Kokrajhar	—	—	—	1	1	2	2	3	4	6
Dhubri	—	—	—	—	—	2	3	3	4	6
NC Hills	—	—	—	—	—	—	—	—	—	2
Hailakandi	—	—	—	—	—	—	—	—	—	3
Darrang	—	—	—	3	6	9	126	129	132	132
Sonitpur	6	12	21	24	63	78	165	177	189	276
N. Lakhimpur	—	3	12	24	42	54	57	63	69	734
Dhemaji	—	—	—	—	3	3	3	28	46	139
Cachar	—	—	4	3	7	15	30	40	77	102
Total	657	1,543	3,185	4,778	6,300	7,442	9,155	10,018	15,647	19,494

Source: Baruah, Sarma, Deka: 1999.

Table 14.2: Size of Small Tea Holdings

Districts	< 5 *Bigha*	5-9 *Bigha*	10-19 *Bigha*	20-29 *Bigha*	30-39 *Bigha*	40-49 *Bigha*	50-59 *Bigha*	60-69 *Bigha*	70 *Bigha* & above	Esti. Total
Tinsukia	241	362	1,085	302	121	215	87	144	38	2,595
Dibrugarh	1,212	2,211	2,020	808	337	166	81	21	11	6,867
Sibsagar	123	613	1,104	429	307	61	45	43	35	2,760
Jorhat	159	1,190	1,571	79	130	26	11	5	3	3,174
Golaghat	102	316	468	606	137	269	179	30	41	2,148
Karbi Anglong	—	—	16	31	7	34	3	6	5	102
Nagaon	13	6	10	12	30	9	12	1	—	93
Marigaon	0	5	8	6	3	—	—	—	—	22
Kamrup	—	12	15	13	11	2	7	17	—	77
Nalbari	—	—	—	3	5	—	—	2	—	10
Goalpara	—	—	122	50	74	—	—	—	—	246
Kokrajhar	—	—	—	4	—	—	—	—	2	6
Dhubri	—	—	—	—	—	—	2	3	1	6
NC Hills	—	—	—	—	1	—	1	—	—	2
Hailakandi	—	—	—	—	1	—	2	—	—	3
Darrang	37	51	22	15	3	4	—	—	—	132
Sonitpur	15	46	123	35	21	12	6	12	6	276
N. Lakhimpur	73	138	152	326	25	13	3	1	1	734
Dhemaji	17	34	31	28	13	3	11	2	—	139
Cachar	12	13	11	15	15	29	3	3	1	102
Total	2,004	4,997	6,760	2,762	1,241	843	453	290	144	19,494

Source: Baruah, Sarma, Deka: 1999.

Table 14.3: Estimated Area Under Small Tea Garden (1999)

Districts	Estimated total area (ha)
Tinsukia	6,574
Dibrugarh	11,839
Sibsagar	6,385
Jorhat	4,687
Golaghat	7,342
Karbi Anglong	439
Nagaon	404
Marigaon	47
Kamrup	391
Nalbari	47
Goalpara	688
Kokrajhar	51
Dhubri	95
NC Hills	10
Hailakandi	15
Darrang	503
Sonitpur	874
N. Lakhimpur	1,620
Dhemaji	433
Cachar	340
Total	42,774

Source: Baruah, Sarma, Deka: 1999.

The lure of tea garden ownership has uprooted various traditional varieties of horticultural crops, e.g., sugarcane in Golaghat, bamboo in Sibsagar and Jorhat, orange in Dibrugarh and Tinsukia and pineapple in Karbi Anglong district. Arecanut has been the common casualty in all these concerned districts. Another plantation called *chomlsum* (*Persea bombycina*) which is used for rearing *muga* larvae has been replaced with small tea gardens in various places.

Causes for Shift

In fact, a large majority has started small tea gardens realising that it is a profitable business option. Availability of unused land in close vicinity helped them to realise their dreams. While there are many who have taken to small tea cultivation as a self-employment option, there are others who undertook it simultaneously with their main occupation. While other horticultural crops are attacked by animals specially elephant, monkeys, etc. tea cultivation remains undisturbed from these attacks. Above all an underlying urge to become a tea planter plays an important part considering the status attached with tea garden owners in Assam.

Production

Production of green leaf from small tea gardens has been estimated at 318.77 million kg during 1999 which has been 15 per cent of total tea production in Assam. In the five major tea growing districts of Upper Assam the yield per hectare varied between 16,000 kg (Tinsukia) to 14,500 kg (Golaghat) whereas in other districts it has been in the range of 10,000 kg to 12,000 kg per hectare respectively.

Price and Profit

The price of green leaf varied widely over the years. In the 5 major tea growing districts of Assam, the average price per kg of green leaf has been Rs. 3.60 during 1990 (where Golaghat had the highest price Rs. 3.90 and Jorhat the lowest at Rs. 3.40). During 1999, the average price has been Rs. 11.45 in the same districts (where Golaghat had the highest price Rs. 12.00 and lowest at Tinsukia Rs. 10.75).

It is estimated that other things remaining constant, the cost of establishing these small gardens are compensated during the fifth year and thereafter on an average profit of Rs. 70,000 to Rs. 80,000 per hectare can be earned by the small tea cultivators.

Floriculture in Central Assam[7]

Floriculture in Assam has been limited to a cluster of villages (namely Bor Dadhi, Khudra Dadhi, Hatdola, Bagta, Ram Dia, Checha, Dodhial Pam, Kismat Bonshar, etc.) in and around Hajo region of Kamrup district. Land grants were given to the people belonging to the 'Malakar' caste who as traditional flower growers started cultivation in these areas. Till date majority of the flower cultivators and persons involved in this trade belong to this particular caste (although people from other castes have also joined this trade over the years) and are settled in Hajo Block.

Among this cluster of villages, the largest village named Bor Dadhi has been surveyed and the changes in flower cultivation over the years has been analysed in this section.

Area under Floriculture

During 1987 floriculture has been limited both in terms of households and area involved. Total households involved were 17 and they cultivated flower in an area of about 50 *kathas*. During the next five years another 7 households started flower cultivation and total land involved doubled to 105 *kathas*. During the next five years another 7 households got involved in floriculture and the area increased substantially to 230 *kathas*. During the ten years period (1987-97) land area under floriculture increased by 21.8 per cent whereas cultivators involved increased by 23.3 per cent respectively.

Floricultural Holdings

There has been several changes in the floricultural holdings during the same period. During 1987, 64.7 per cent of the holdings were below 3 *kathas* where the largest holding was of 8 *kathas*. Over the next five years, maximum number (54.2%) of the holdings ranged between 3 to 6 *kathas* and the largest holding ranged between 9 to 12 *kathas*. During 1997, majority (23.3%) of the holdings were in the range of 9 to 12 *kathas* and 10 per cent of the total holdings were above 15 *kathas*. Therefore, over the years the size of floricultural holdings increased as more cultivators and land got involved in flower cultivation (See Tables 14.4 and 14.5).

Table 14.4: Floricultural Holdings, 1987

(*in Katha*)

Habitation	< 3		3-< 6		6-< 9		Total	
	Total	PC	Total	PC	Total	PC	Total	PC
Bamunbari	—	—	1	100	—	—	1	100
Bohol	2	100	—	—	—	—	2	100
Diyar Bori	8	72.7	2	18.2	1	9.1	11	100
Mali	—	—	1	100	—	—	1	100
Na Basti	1	100	—	—	—	—	1	100
Patoni	—	—	1	100	—	—	1	100
Total	11	64.7	5	29.4	1	5.9	17	100

Source: Chakraborty, 1999.

Land-Use Prior to Floriculture

The holdings which are used for floriculture has been largely used for growing other crops. In 44 per cent of these holdings there has been crop replacement. Crops like sugarcane, potato, tomato, brinjal, mustard and jute has been replaced by flower cultivation. In another 46 per cent of the floricultural holdings, *Kothia* (paddy seedlings) are grown other than flower cultivation. Only 10 per cent of the floricultural holdings are in those areas which were barren uncultivated land prior to flower cultivation.

Causes for Shift

All the flower cultivators cited profit to be the major reason for their shift towards floriculture. Those who cultivated sugarcane, brinjal, tomato, potato, etc. stated that their crops were often attacked by the monkeys which wrecked havoc ultimately resulting in huge losses. Once they shifted to flower cultivation the 'monkey menace' have stopped altogether. Moreover those involved in jute cultivation cited high labour cost involved with jute other than unremunerative price for the same and, therefore, shifted to flower cultivation which by and large is manageable with family labour.

Floral Season and Variety

Floriculture is a seasonal affair in Assam. The floral season starts during the last week of October or early November and continues upto March or early April. About 50 per cent of the cultivators stated their floral season to be of 120 days and 40 per cent stated it to be of 90 days.

Floriculture lacks variety. Only Marigold (*Narji*) is grown commercially. Although there has been experiments with other varieties, namely *Rajani Gandha*, Gladious, Rose, etc. they did not prove to be commercially viable. Nearly 74 per cent of the cultivators mentioned that they suffered from information gap about other floral varieties and proper method of their cultivation.

Production of Flowers

Nearly 50 per cent of the flower cultivators had production of 6 lakh

Table 14.5: Floricultural Holdings, 1997

(*in Katha*)

Habitation	< 3 katha		3-< 6		6-< 9		9-<12		12-< 15		15 and above		Total	
	Total	PC	Total	PC	Total	PC	Total	PC	Total	PC	Total	PC	Total	PC
Aliakh	—	—	—	—	1	50	1	50	—	—	—	—	2	100
Bamunbari	—	—	—	—	—	1	1	100	—	—	—	—	1	100
Bohol	1	20	4	80	—	1	—	—	—	—	—	—	5	100
Diyar Bori	1	5.6	8	44.4	3	16.7	3	16.7	1	5.6	2	11.1	18	100
Mali	—	—	—	—	—	—	1	50	—	—	1	50	2	100
Na Basti	—	—	—	—	—	—	—	—	1	100	—	—	1	100
Patoni	—	—	—	—	—	—	1	100	—	—	—	—	1	100
Total	2	6.7	12	40	4	13..3	7	23.3	2	6.7	3	10	30	100

Source: Chakraborty, 1999.

Table 14.6: Production of Loose Flowers

(*in Katha*)

Land production	Upto 1 lakh		Upto 3 lakh		Upto 6 lakh		Upto 9 lakh		Upto 12 lakh		Upto 15 lakh		Total	
	Total	PC	Total	PC	Total	PC	Total	PC	Total	PC	Total	PC	Total	PC
< 1	1	50	1	50	—	—	—	—	—	—	—	—	2	100
3-< 6			1	8.3	8	66.7	1	8.3	1	8.3	1	8.3	12	100
6-< 9					3	75	1	25					4	100
9-< 12					3	42.9	4	57.1					7	100
12-< 15							1	50			1	50	2	100
15 above					1	33.3			2	66.7		3	100	
Total	1	3.3	2	6.7	15	50	7	23.3	3	10	2	6.7	30	100

Source: Chakraborty, 1999.

loose flowers during the season and another 23 per cent of them produced upto 9 lakh loose flowers during the same period. Only 6.7 per cent of the cultivators could produce upto 12 lakh flowers (See Table 14.6).

Income from Floriculture

Income from floriculture has been largely underestimated during reporting by the flower cultivators. (A threat of tax on flower cultivation loom large over their head due to some unforeseen reasons). About 33 per cent of the cultivators earned between Rs. 7,000 and Rs. 15,000, only 10 per cent earned more than Rs. 15,000. For the rest income ranged between Rs. 1,000 to Rs. 7,000 respectively.

However after detailed information and enquiry it has been estimated that 1 *bigha* of floricultural land can fetch an average income of Rs. 25,000 in a season. This estimate may vary within a specific range due to differences in input costs, man-days utilised and above all fertility of the soil.

Iri Paddy Cultivation in *Char* Areas

There are 2,251 *char* villages in Assam spread over 14 districts along the route of the mighty Brahmaputra. The *chars* or the riverine tracts are a special feature of this river. They have a unique geo-morphology of their own. These remote tracts were brought under cultivation by migrating peasants from East Bengal (present day Bangladesh) during the last part of nineteenth century. Their migration was favoured and facilitated by the Britishers as a part of the colonial plan to increase cultivable land (specially cultivation of cash crops) so that not only more foodgrain is produced but also enough revenue is generated in the State. The *char* dwellers introduced large scale jute cultivation[8] which till date is one of the major crops of Assam. They diversified and commercialised agriculture in the State by bringing in large tracts of uninhabited, uncultivated and remote areas under cultivation including the *char* areas. By dint of hard work and gifted with superior agricultural skill, these *char* dwellers reaped gold from the barren land. This process continues till today. However despite their toil, they are among the most poorest, illiterate and neglected population groups of Assam.[9]

Household survey has been undertaken at Nishaner *char* village in Mondia Block of Barpeta district and certain new trends concerning changes in agricultural practices were observed.

From the latter half of the 90's a new variety of paddy known as *iri* has been introduced in the *char* areas. This cultivation got boosted with the availability of shallow water pumps either as a part of the Government scheme or personal purchase. It starts during *Magh/ Phalgun* and is harvested after 3 months. During this period there is negligible rainfall which necessitates the use of shallow water pumps. This variety of paddy is undertaken with HYV seeds, high doses of chemical fertilisers, employing large number of agricultural labour and substantial amount of water. As the yield is higher, the availability of foodgrains has, therefore, increased for those who undertake *iri* cultivation. This has reduced their economic vulnerability to a great extent.

Another important trend noticeable has been large scale leasing of land in *kayam* (settled) areas by *char* dwellers for *iri* cultivation. These cultivators lease land in nearby areas, e.g., Kalobari, Tupamari, Jamlai, Piju Para, Maloncha, etc. If their economic condition improve after couple of harvests, they aspire to purchase cultivable land in these areas. In the following section certain specific details associated with this cultivation are analysed.

Area under Iri Cultivation

This pattern of cultivation is practiced both in *char* and *kayam* areas. Cultivators of the *char* areas, either lease land or purchase land for cultivation in *kayam* areas. During 2003, cultivators from this *char* village involved 62 *bighas* of land under *iri* cultivation out of which about 60 per cent were in *kayam* areas and the rest in *char* areas. Moreover 40 per cent of the cultivators leased land for *iri* cultivation. Although the exact amount of land under this cultivation during previous years could not be ascertained yet it was clear during the course of survey that it is increasingly becoming popular among the *char* dwellers.

Size of Holdings

About 45 per cent of the holdings were in the range of 3 to 5 *bighas*, more than 24 per cent of the holdings ranged upto 2 *bighas* while 31 per cent of the holdings were above 5 *bighas*.

Causes for Growing Iri

Other than their enterprising nature those cultivating *iri* had certain common features, i.e., either they cultivate *iri* who have lost large tracts of their cultivable land due to land erosion (a phenomenon quite common in *char* areas of Assam) or it has been undertaken by those cultivators who have a ready flow of cash—either they themselves or anyone of their family members are service holders or engaged in business who provides with both cash for regular supply of costly inputs required in this cultivation.

In all the varieties the requirement for more paddy to sustain their family all throughout the year has been another cause for undertaking *iri* cultivation. In all the households involved in this cultivation, average members were more than 10 per household which required large amount of foodgrain. So in order to reduce their market dependence on paddy, they undertook *iri* cultivation.

Terms of Leasing

Those cultivators who are unable to cultivate in their own land generally lease land in *kayam* areas. About 40 per cent of the cultivators leased land for *iri* cultivation and among them 70 per cent leased in *kayam* areas. Terms of leasing were generally in kind. Only 1 household leased-in land in cash terms at Rs. 5,000 per *bigha*. But this lease has been annual in nature where after *iri* cultivation, the cultivators cultivated other crops also.

.Where rent is paid in kind, it is higher in *kayam* than in *char* areas. If the shallow pump is leased along with land then average rent is more than 7 *maunds* of paddy per *bigha* whereas in case of land it is 4 *maunds* per *bigha* respectively.

Production and Income

The overall productivity of *iri* paddy is more than all other varieties (*Aush*/*Aman*, etc.) of paddy cultivated by the *char* dwellers. But the cost of production of this variety is higher than rest due to higher input costs, e.g., cost of owning or leasing shallow machine, diesel/

kerosene, chemical fertilisers, employing large number of agricultural labour, etc.

Average production is 17.9 m/*bigha*. It is higher in *char* areas (19.8 m/*bigha*) than in *kayam* (16.7 m/*bigha*). Moreover productivity of leased land (16.2 m/*bigha*) is lower than self-cultivating land (18.9 m/*bigha*).

Iri variety of paddy is coarse compared to other varieties of paddy grown by *char* dwellers. So the *iri* cultivators use this produce for self-consumption and convert the excess paddy from *aush* and *aman* cultivation into marketable surplus. Only 2 cultivators sold paddy among all the households undertaking *iri* cultivation. Actually both these households were comparatively well-off, owning 20 *bighas* of land each, having family members in various profession and business which provided them the scope to convert their *aush* variety of paddy into marketable surplus and earn ready cash accordingly.

EMERGING TRENDS

Changes in agriculture is a welcome phenomenon as long as it does not reduce the foodgrain availability or its entitlement for the population group which undertake it. In case of Assam, there are no apparent signs of the above mentioned phenomenon. Therefore, changes in agricultural practices as a whole can be regarded as a welcome phenomenon. Nevertheless in all the three crop varieties there are certain trends which are analysed in this section:

(i) Although the concept of small tea garden is new yet Upper Assam has been familiar with tea plantation and processing for over last 150 years. So the availability of know-how for tea cultivation, maintenance of tea gardens, specialised labour force, infrastructure, etc. were already present which has been accommodated by the emerging and enterprising section of the middle class to suit their (small) scale of operation. This has certainly changed the face of barren land in Upper Assam and has also generated employment and income for 1.5 lakh people of this region.

Similarly in case of floriculture, the *Malakars* possessed indigenous skills related with flower cultivation which has been learnt by other villagers in the cluster of flower growing villages. This has helped to bring in more areas under flower cultivation. Of course, floriculture now is no more limited to *Malakars* only, cultivators belonging to other castes have also started flower cultivation.

In case of *iri* cultivation in *char* areas, although the variety happens to be new and availability of shallow water pumps has helped this process yet paddy cultivation has been an age old practice of the skilful cultivators in these areas which facilitated the easy accommodation of this newer variety.

So in all the three cases, existence of indigenous skill of the cultivators regarding crop production and management has been the benchmark which have helped the process of change and in no case a completely new crop (which was non-existent in the region) has replaced the existing varieties.

(ii) Production of tea and flower have succeeded in increasing the income of the cultivators as both the produce were sold as marketable surplus. *Iri* cultivation has undoubtedly food availability and entitlements in *char* areas. However, *iri* produce is always kept for self-consumption and surplus (if any) from cultivation of *aush* or *aman* varieties of paddy, are only used as marketable surplus.

(iii) In all the three cases, availability of regular flow of cash has been an important factor which helped the process of change. In tea cultivation, majority of finance came from family members or friends, in case of flower cultivation it has been self-employment and business whereas in case of *iri* cultivation it has been a combination of service and business activities either by the cultivators, themselves, or any of their family members. Formal credit therefore has been insignificant in all the cases citing changes in cropping pattern.

(iv) There has been crop replacement in all the three forms of cultivation but the variety of those replaced by small tea growers has been maximum. Bamboo, orange, pineapple, sugarcane, arecanut, som (*Persea bombycina* used for *muga* cultivation) has been replaced by tea cultivation. This should be taken into consideration as it might affect the crop diversity of this region in the long run.

(v) The overall socio-economic impact among all the three crops analysed here has been maximum in case of small tea cultivation. The phenomenal rise in the number of small tea growers has created large number of employment opportunities. Moreover there has also been number of leaf factories and tea nurseries in Upper Assam which has given further momentum to the process of employment and income generation in this region.

(vi) Those involved in small tea gardens were economically sound and their aspirations were somewhat middle-class oriented in contrast to those who cultivated flower and *iri*. A survey of the small tea gardens revealed that with their income from tea cultivation, about 5 per cent purchased cultivable land either for paddy (1.5%) or for further tea cultivation (3.5%) whereas 23 per cent purchased either two wheeler (19.4%) or personal car (3.6%). This aspect has been absent in case of both flower and *iri* cultivators of Assam.

Policy farmers may consider these emerging micro-trends while framing policies for sustainable agriculture in Assam.

NOTES AND REFERENCES

1. Subsidies to the agricultural sector has been prevalent in both the developed and the developing nations. But with the enactment of the WTO regime the balance of subsidies has tilted heavily in favour of the developed nations. While the developing countries have reduced subsidies under the pressure of reducing tariff barriers and quantitative restrictions, the developed countries has either maintained or increased there subsidies to their agricultural sectors. Under the guise of 'blue' and 'green' boxes they have continued with huge subsidies while putting subsidies of the developing nations under 'amber' box, they have been forced to reduce them under WTO conditions. Doha and Cancun Rounds of Trade Negotiations bear testimony to these facts.
2. For a detailed analysis on 'changes in rural economies', 'agricultural diversification' and 'cash crops' see Dreze, Jean and Sen, Amartya, *Hunger and Public Action* (New Delhi: Oxford University Press, 1989).
3. This section is based on some selected seminar papers. National Seminar on '*Plantation Economy of the North East with Special Reference to the Small Tea Garden Movement in Assam*', 12-13 November 1999, Guwahati,

sponsored by The Society for Conflict Analysis and Resolution (SOFCAR), New Delhi and organised by the North East Foundation, Guwahati:

(a) Baruah, Swapan; Sarma, Dhrubajit and Deka, Mridul, 'Status of Small Scale Tea Cultivation in Assam'.

(b) Sarma, Dhrubajit; Baruah, Swapan and Bhuyan, Ratna Pratap, 'Prospects and Problems of Small Tea Growers of Assam'.

(c) Borah, Dhruba Jyoti, 'A Survey of the Small Tea Garden Movement of Assam'.

(d) Gogoi, Jayanta Kumar, 'Boom in Indian Tea Industry and Small Tea Growers of Assam'.

4. Tea has been always synonymous with large scale cultivation. The Britishers nipped the bud of small tea cultivation by restricting tea plantation upto a maximum of 71 bushes by small tea growers. This was an imperial law and any violation of it attracted a fine of Rs. 10,000. It was during the first AGP regime during 1985 that the small tea growers were allowed to cultivate in land measuring upto 30 *bighas*.

5. After the Post-emergency Janata regime, the idea of small tea garden and the huge economic prospects attracted with this venture was highlighted by the then Agriculture Minister of Assam, Shri Soneswar Bora.

6. A standard definition of small tea growers is yet to emerge. According to Tea Board a small tea grower's land under tea cultivation should not exceed 75 *bighas*. This norm is also followed by Small Tea Growers Advisory Programme (STAP) of AAU, Jorhat. However, Government of Assam's maximum limit for small tea garden is 30 *bighas*. The All Assam Small Tea Growers Association has requested this limit to be raised to 250 *bighas*.

7. For detailed analysis see Chakraborty, Gorky, *Floriculture in Assam: Trends, Perspectives and Shifts* (Guwahati: OKD Institute of Social Changes and Development, 1999).

8. For a detailed analysis of jute cultivation in Assam and the contribution of the peasants from erstwhile East Bengal see Guha, Amalendu, *Plater-Raj to Swaraj: Freedom Struggle and Electoral Politics in Assam 1829-1947* (New Delhi: ICHR, 1977).

9. The *Char*-dwellers of Assam are among the most neglected population groups of Assam. About 69 per cent of the total families in *char* areas live below poverty line. Their literacy rate is pathetic at 19.31 per cent. Land erosion is a recurring phenomenon. The State seems to be apathetic towards the inhuman condition of the *char* dwellers. For details see *Socio Economic Survey Report of Char Area of Assam, 2003* (Dispur: Directorate of *Char* Area Development, 2003).

10. For details regarding the Guwahati flower market, see Chakraborty, Gorky, 'Where Flowers Degenerate man Cannot Live', in *Journal of North-East India Council for Social Science Research*, Vol. 24, No. 1, April, 2000.

15

AGRICULTURE AND SUSTAINABLE DEVELOPMENT
A Case Study of Assam

— B.P. SAHU

North-Eastern Region with natural endowment of forests, orchards and wide ranging field crops and other flora, offers immense potentialities for the growth and development of agriculture. The region is characterised by diverse geological, physiographic and agro-climatic features which makes the region unique. The North-Eastern Region (NER) occupies a geographical area of 2.74 lakh square kilometres. Physiographically, the region is divided into four units, namely: Eastern Himalayas; Eastern Mountains; Meghalaya—Mikir tablelands; and the Brahmaputra valley.

This in turn has given rise to six agro-climatic zones which form the primary basis for development of agriculture in the region. The broad parameters governing the mosaic pattern of agro climates comprise rainfall—ranging between 2200 mm to 4500 mm, temperature between 10-35 degree Celsius, relative humidity 45-95% per cent, *p*H 5.5 to 7.00, varying altitude, longitude, etc. The region characterised by a wide spectrum of agro climates ranging from tropical humid to alpine temperature and rich in bio genetic resources, is congenial for multifarious agricultural development. The entire agricultural development revolves around this chequered pattern. North-East India sustains two-third of the country's forests. Also two-thirds of its area is under cultivation. But surprising enough the region is not self-sufficient in foodgrains. While it produces in excess of other type of cash crops like tea, ginger and potatoes. In

1998-99, the production of potato was 9,97,000 tonnes grown in 114.5 thousand hectares of land compared to the all India production rate of 2,24,95,000 tonnes. The total share was 4.34 per cent of the national production. The yield per hectare is 85,33,000 kg against the all India total of 1,75,74,000 kg per hectare. The yield from Tripura alone is 1,76,00,000 kg per hectare. The region produced 1,07,000 tonnes of ginger in 18,000 hectares of land. The yield on ginger is 5,944.00 kg per hectare. In terms of horticulture production, the region stands out in terms of bananas and pineapples. In the year 1998-99, the area under banana was 60,000 hectares and production was 7,36,000 tonnes. The yield stood at 12,267.00 kg per hectare. The region produced 4,86,000 tonnes of pineapple against all India figures of 10,71,000 tonnes.

The North-East has nearly about 30 per cent of the total national water resources. The region has ground water potential of 25,320 M Cubic metre. This apart, existence of a large number of rivers (19,198 km), reservoirs (8.123 ha), streams, natural lakes/beels (14,37,000 ha), ponds, and tanks (41,624 ha) other estuaries open up opportunities for extensive agricultural development. Stretched over 1.54 lakh square km, the region covers nearly 55 per cent of the geographical area with forest resources consisting of 650 valuable forest species. It has low land paddy fields of 94,80,800 ha. While rest of the country utilises 90 per cent of the water in agriculture sector, the use of water for agriculture and allied activities in NER are negligible. A major reason for the non-utilisation of water is the frequent flood in the region. Four North-Eastern States witness flood every year, Assam being the most acute where 8,000 hectares of fertile land along the Brahmaputra river are eroded every year. Since 1954, nearly 4 lakh hectares of fertile agricultural land accounting for seven per cent of the plains in Assam covering 18 districts have been eroded.

However the region is heavily dependent on State like Andhra Pradesh, Uttar Pradesh and Punjab for its supply of wheat and rice. The region does not produce sufficient foodgrains for its consumption. The region spends over a thousand crore of rupees in importing cereals. Assam alone accounts for 9.56 lakh MT annual deficit of rice production.

Emerging Agricultural Scenario in Assam

Topographical Features

Assam comprises an area of 78,523 square kilometres with 25 districts. Except for the districts of Karbianglong and North Cachar hills, Assam is generally composed of plains and river valleys. It is divided into three principal geographical regions; the Brahmaputra valley in the north; the Barak valley in the south; and the Mikir (Karbi Anglong) and Cachar hills that divide the two regions. The elongated valley of the mighty river Brahmaputra wholly occupies the northern part of Assam. Most of Assam's population live in this valley. The foothills of the Himalayas bound the Brahmaputra valley to the north and another lower range hills and mountains to the south. In the centre part of Assam, to the south of the hills is the Barak valley, which is contiguous with Bangladesh. The Brahmaputra valley is the dominant physical feature of Assam. The Brahmaputra enters Assam near Sadiya at the extreme north-east corner and runs westward for nearly 450 miles before turning south to enter the plains of Bangladesh. The river valley, rarely more than 50 miles wide, is studded with numerous low, isolated hills and ridges that abruptly rise from the plain. The valley is surrounded on all sides, except the west, by mountains and is intersected by many streams and rivulets that flow from the neighbouring hills to empty into the Brahmaputra. Assam lies in the low productivity-high potential zone, where the productivity is low despite abundant water availability and good soil.

Resources and the Sectoral Contribution

Assam is a State rich in natural resource like oil and natural gas, coal, rubber, tea and some mineral like granite, limestone, and kaolin. The present State is much smaller than what it was forty years ago. It is still the largest economy of NER. It is like the gateway of other North-Eastern States with an infrastructure index (according to CMIE in 1992-93) of 93, quite close to the average of 100. Although it is more industrially developed than the other North-Eastern States, it is primarily an agrarian economy with 74 per cent of its population engaged in agriculture and allied activities. The State is known for its large industries in the petroleum sectors. In

Table 15.1: NSD Product at Factor Cost by Industry of Origin at Constant (1993-94) Prices

(*Rs. in lakhs*)

Industry	1993-94	1997-98	1998-99	1999-2000	2000-01 (*Q*)	CARG (%)
Agriculture	5,02,168	5,47,705	5,19,522	5,49,722	5,64,894	1.70
Sub Total Primary Sector	6,47,448	7,03,174	6,74,218	7,04,383	7,23,595	1.60
Total NSDP	13,47,683	14,70,357	14,57,444	15,58,037	16,23,618	2.70
India	6,85,91,200	8,90,71,200	9,48,98,200	10,11,22,400	—	6.68

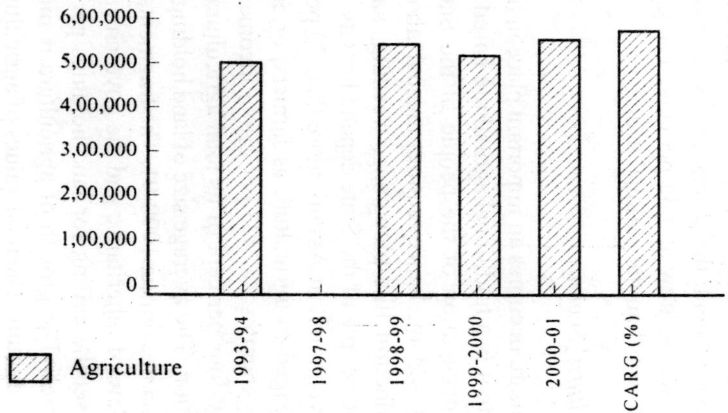

Fig. 15.1: Contribution of Agriculture to NSD

Source: Directorate of Economics and Statistics, Government of Assam, 2002.

1999-2000, the contribution of the tertiary sector to the State Domestic Product (SDP) is the highest (39.46). Over the years, there has been a shift in the sectoral contribution from primary to the tertiary sector while the secondary sector indicates a fluctuating contribution to the SDP. Assam's Net State Domestic Product at constant (1993-94) prices was Rs. 16,23,618 lakh in 2000-01. During 1993-94 to 2000-01 the NSDP at constant price grew by an average rate of 2.70 per cent per annum. The growth rate was lower than the all India rate during the same period. During the period 1993-94 to 2000-01, the compounded annual average rate of growth of the primary sector was 1.60 per cent; the secondary sector was 4.58 per cent while the tertiary sector grew at 3.32 per cent. The growth rate achieved was much lower than the national average.

Agricultural Scenario

Agriculture occupies an important place in the economy of the State. Till date it is a traditional agriculture based economy and forms the major occupation of the people of the State. It is relatively non-mechanised and hence requires manual labour during various stages of production like ploughing, weeding, harvesting and threshing, etc. The people of the State depend heavily on agriculture sector for their livelihood. In Assam, more than 70 per cent of the population are engaged in agriculture as farmers, or as agricultural labourers, which generates Rs. 423 per capita income from the sector (1990). Seventy-four per cent of its total agricultural area is devoted to rice cultivation. The average size of land holding per household was only 1.30 hectares during 1990-91. The fragmentation of land is attributed to the law of inheritance and the government land reform measures which set the ceiling for land holdings per family to 50 *bighas* or 16.6 acres. The growth of agriculture is miserably low at 1.87 per cent per annum. The total values of agricultural crops are growing at the rate of 2.14 per cent per annum. The total foodgrain production in 1999-2000 was 4.04 million tonnes and it occupies about 70 per cent of the gross cropped area in the State. Main cash crops are jute, tea, cotton, oilseeds, sugarcane, fruits and potatoes. Wheat production in 1999-2000 was 0.1 million tonnes, rice was 3.9 million tonnes and pulses were 64.688 tonnes. Although rice is the major crop of Assam, its productivity over the years has not increased while the other crops have slightly increased in both productivity and land acreage. For example, while rice yield per hectare in 1970-71 was 1,022 kgs per hectare, it was 1,261 kgs in 1990-91. The yield of wheat was 583 kgs in 1970-71 and it increased to three-fold from 583 kgs per hectare to 1,455 kgs in 1990-91. A similar increase was observed in jute, sugarcane, potato, and mustard.

Population and Agriculture

While the economy of the State is ailing for all these years, the population growth is exceptionally high in the State. While the population of the country during the 1961-91 decade grew by 92 per cent, it has grown by 107.4 per cent in Assam during the same period. Population density in Assam was 286 per square kilometre as against the national average of 267 per square kilometre in 1991. As per 2001 census, the population of the State is staggering 2,24,14,322 with a growth rate of 24.2 per cent and by 2025, it would reach around 4 crore which would put tremendous pressure on land, infrastructure and its people. The population density stands at 340 people per square kilometre as per 2001 census.

As seen from Table 15.2, there was an increase in the number of cultivators from 35,59,117 in 1991 to 37,41,912 in 2001, i.e., an increase of 0.05 per cent. The number of males engaged as cultivators declined by 7.2 per cent whereas the female cultivators increased by 56 per cent during the decade under reference. As regards agricultural labourers, the total increase percentage hovered around 47 per cent with 24.44 per cent males and 189 per cent for females during the decade from 1991-2001.

As seen the shifting of the trend is towards females as agricultural labourers and cultivators. It is also evident that the males have diverted from the primary sector to secondary or the tertiary sector for their livelihood and sustenance.

Land Pattern in Assam

Figure 15.2 reveals that the net area for cultivation increased by only 0.7 per cent while the agricultural land increased by only 0.64 per cent from 1996-99. The figure also indicates that there was a steady growth of uncultivable land to the extent of 10,852 hectares during the period of reference. The area sown more than once showed a decline trend to the tune of 4,886 hectares from 1996-99. The net

Table 15.2: Decadal Population engaged in Agriculture

	Population			Cultivators			Agriculture labourers		
	Total	Male	Female	Total	Male	Female	Total	Male	Female
1991	2,24,14,322	1,16,57,989	1,07,56,333	35,59,117	28,70,583	6,88,534	8,44,964	6,82,573	1,52,391
2001	2,66,38,407	1,37,87,799	1,28,50,608	37,41,912	26,61,619	10,80,293	12,89,902	8,49,434	4,40,468

Source: Census Report, 2001.

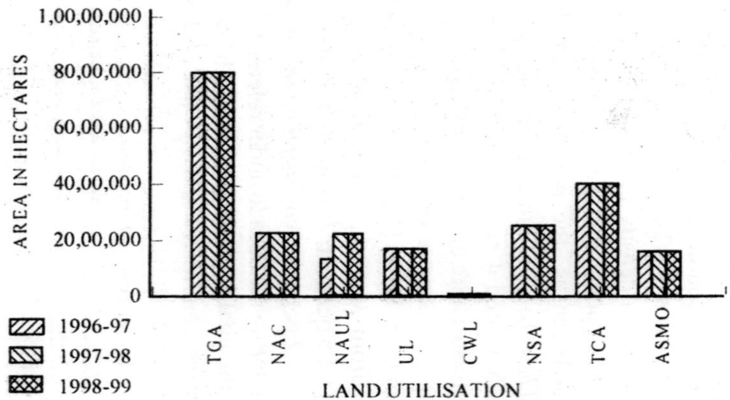

Fig. 15.2: Classification of Area in Assam

Source: Directorate of Economics and Statistics, Government of Assam 2002.

sown area was 68.54 per cent to that of the total cropped area as of data available in 1998-99. Assam accounted for 5.72 lakh gross irrigated area in 1971-72 and there has been no addition of a single hectare till 1991-92. The irrigated area as proportion to cropped area in 1971-72 was 19.8 per cent, in 1981-82 was 16.5 per cent and in 1991-92 was 14.8 per cent thus showing a decline trend over the decades. The gross irrigated area in 1991-92 accounted to 5,72,000 hectares.

Cropping Area

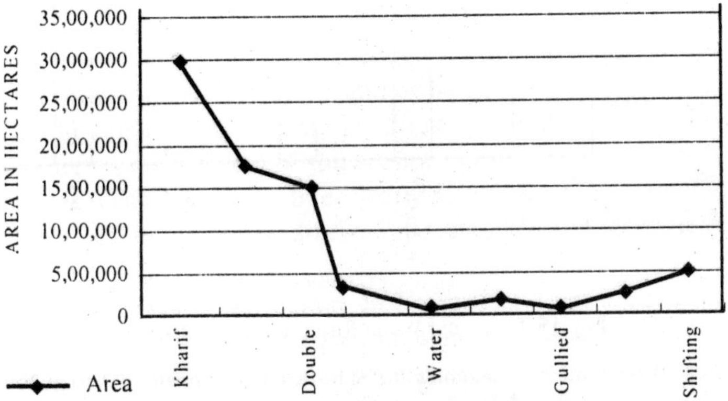

Fig. 15.3: Land Utilisation Pattern of Different Crops in Assam

Source: Directorate of Economics and Statistics, Government of Assam, 2002.

Figure 15.3 shows the land utilisation pattern of different crops in Assam. The four measure areas of utilisation are confined to *Kharif*, *Rabi*, double cropping and shifting cultivation. The area utilised are to the extent of 2,95,824 hectares for *Kharif*, 17,44,511 hectares for *Rabi*, 14,87,041 hectares for double crop and 4,82,571 hectares for shifting cultivation. As seen from the figure the other left out land are waterlogged, swampy, gullied and also covered by grass.

Water Resources

Brahmaputra and Barak rivers alone constitute 32 per cent of the country's water resources and there is very less utilisation of Brahmaputra's water in the region. It is well known that Brahmaputra has surplus water between May and October. The inadequacy of an irrigation network has been a major impediment in improving the output of the region. Owing to paucity of funds, the non-tapping of ground water resources is inexplicable. Assam has appreciable ground water reserves, which have not been fully utilised. The abundance of rains ensures regeneration of this resource as yet its utilisation has remained at a very low level.

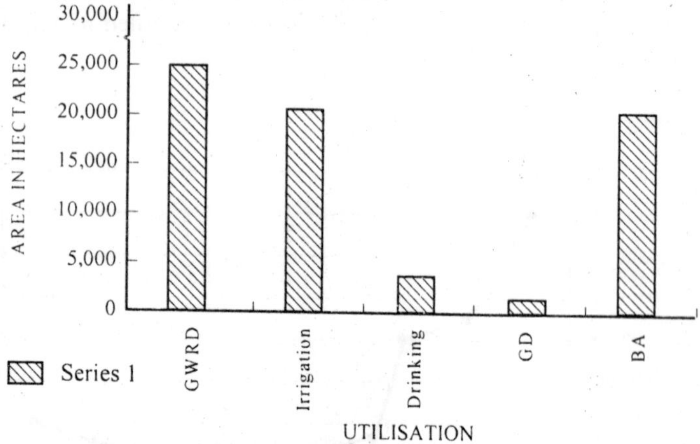

Fig. 15.4: Ground Water Resource and Utilisation

Source: Directorate of Economics and Statistics, Government of Assam 2002.

Assam accounted for 5.72 lakh gross irrigated area in 1971-72 and there has been no addition of a single hectare till 1991-92. The irrigated area as proportion to cropped area in 1971-72 was 19.8 per cent, in 1981-82 was 16.5 per cent, and in 1991-92 was 14.8 per cent thus showing a decline trend over the decades. The gross irrigated area in 1991-92 accounted to 5,72,000 hectares. The figures shows the ground water resources available in the State. If utilised and harnessed properly, it can bring a new revolution in the agricultural sector.

Foodgrain Production in Assam

Rice is the major crop claiming the lion's share in the area, water and budgetary allocations, besides being the dominant staple in the dietary habits of the people. This is evident from the fact that rice, as a major crop is grown thrice a year. However the production pattern shows no subsistence increase from 1993 to 2000. Next to rice other cereals

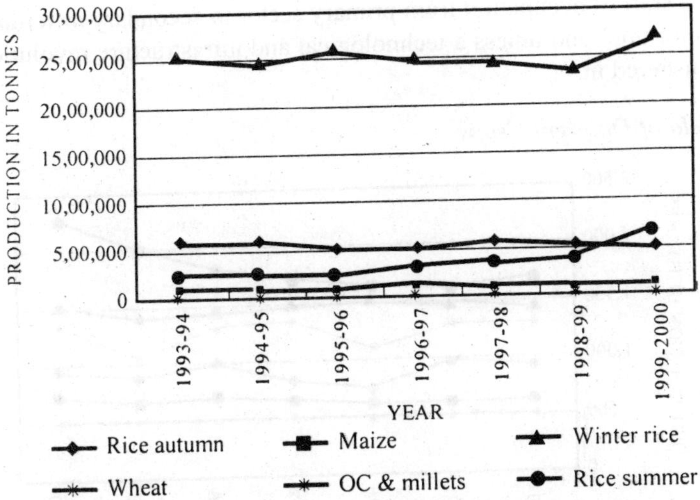

Fig. 15.5: Production of Crops

Source: Directorate of Economics and Statistics, Government of Assam, 2002.

and millets are grown which only showed a substantial increase in the year 1995-96 after which the production stagnated and was around 5,400 tonnes up to the year 1999-2000. In case of maize the production showed a marginal increase over the period from 1993-1998 after which the production declined in the year 1998-2000. The slow growth of crops can be attributed to inadequate production technology, price variability, production risk and low level of irrigation that lead to stagnation yield of these crops.

Production of rice is almost hovering around 1,324 to 1,548 kg per hectare with an annual growth rate of 2.24 per cent in the period between 1993-2000. The cause is attributed to the constraints like cultivation on rain fed, marginal and sub-marginal lands, high susceptibility to insects, pests and diseases, weather aberrations (including flood) and lack of genetic breakthrough. Consequently OC and millets, which was 4,534 tonnes during 1993 registered a growth rate of 32.37 per cent in 1995-96 and from 1995-96 showed a stagnated production around 5,429 tonnes in average up to the year 1999-2000.

As evident from Fig. 15.5, increase in production were smail and future increases are expected to be less than moderate because of a shift of occupation from primary sector to secondary or tertiary sector until and unless a technological and infrastructure revolution is ushered in.

Yield of Different Crops

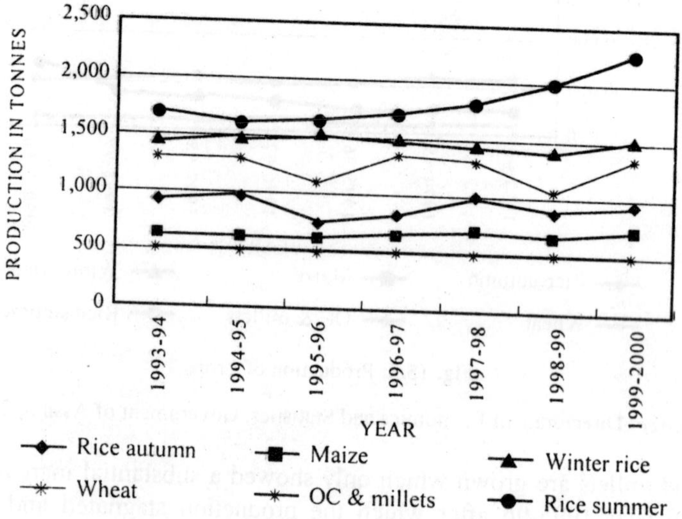

Fig. 15.6: Average Yield of Crops

Source: Directorate of Economics and Statistics, Government of Assam, 2002.

The average yield of rice (autumn) was not substantial as in the case of summer rice. Rice (summer) showed an increase of 4.63 per cent yield over a period of seven years from 1993-2000, whereas in case of winter rice there was no substantial increase of yield and it almost stagnated at 1,415 kg per hectare. There was a very meagre impact of technology on production levels of these crops due to marginalisation factors. Other crops like maize and autumn rice showed no absolute increase in production over the reference period from 1993-2000.

Area of Different Crops

The area under rice recorded a poor growth rate of only 0.60 per cent during the period between 1993-2000. The trend also shows diversification of land to summer cultivation from wheat. The area

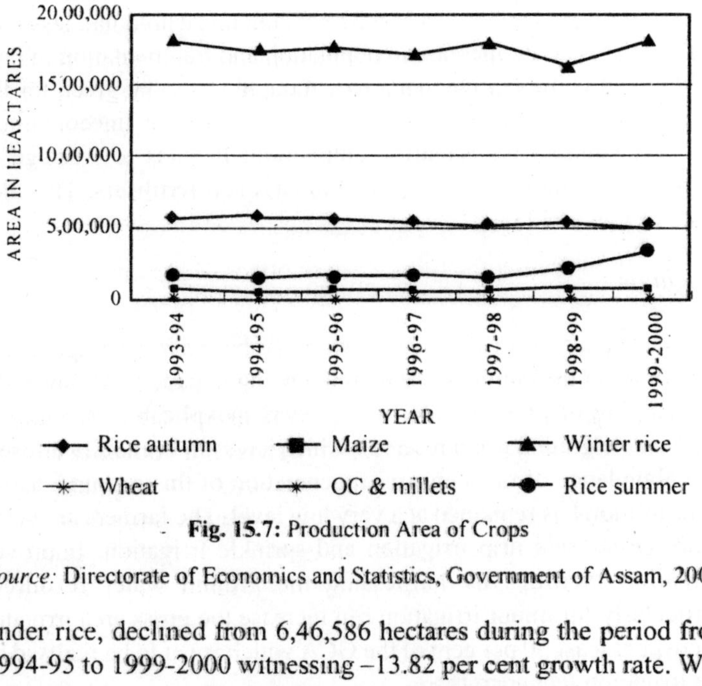

Fig. 15.7: Production Area of Crops

Source: Directorate of Economics and Statistics, Government of Assam, 2002.

under rice, declined from 6,46,586 hectares during the period from 1994-95 to 1999-2000 witnessing –13.82 per cent growth rate. With reduction in area there is decrease in production from 6,19,126 tonnes to 5,14,156 tonnes during the reference period. During 1993-95 production however increased from 5,86,620 tonnes to 6,19,126 tonnes showing a growth rate of 5.54 per cent. After 1995, there was a reduction in the production level till the year 2000 as compared to 1994-95. Summer rice in terms of area exhibited a growth rate of 129.86 per cent from 1993-2000 whereas winter rice witnessed –0.21 per cent growth rate during the same period.

Problems of Agricultural Development

Size of Operational Holdings

The size of operational holdings less than two hectares is uneconomical and falls in the category of small and tiny holdings. In Assam the average size of operational holding during the period from 1970-71 to 1990-91 has come down from 1.47 hectares to 1.31 hectares. Small and marginal holdings account for 80 to 90 per cent of the total holdings. In the year 1971-72, 27.12 per cent of the household were landless and 99.22 per cent household had less than 6.07 hectares. With increase in population and fragmentation of land holdings, it is imperative that a new thought has to be given for the land reforms. Such small size of agricultural plots is uneconomical and less suitable for scientific cultivation. It leads to wastage of labour, time and other inputs like biomass and fertilisers. This also leads to low yield rate.

Irrigation

The inadequacy of an irrigation network has been a major impediment in improving the output of the region. Owing to paucity of funds, the non-tapping of ground water resources is inexplicable. Assam has appreciable ground water reserves, which have not been fully utilised. The abundance of rains ensures regeneration of this resource as yet its utilisation has remained at a very low level. The farmers are yet to adopt large-scale drip irrigation and sprinkle irrigation. Improved irrigation facilities by harnessing the ground water resources particularly for minor irrigation can increase the gross area irrigated to cover at least 50 per cent of the GCA which is yet to be realised by the irrigation technocrats.

Flood Damage and Flood Control Measures

Floods over the years have affected the people in terms of damage to property, crop, cattle, and loss of human lives besides shattering the socio-economic fabric of the State. The total area affected in 1969 was 8,607 square kilometres, which was further, increased to 8,607 square kilometres in 1970 and to 3.06 lakh hectare in 1978. The total damage was estimated at 427.02 lakh of rupees in 1978.

Of the 25 districts in the State, 22 districts experienced the worst flood in 50 years in Assam last year. As estimated 48 lakh people are affected by floods, which brought untold damage to the agricultural production (*The Telegraph* 25th July, 2002). Estimated loss is of 2.4 per cent of SDP due to flood damage. In the year 1987-88, 10.49 lakh hectares of crop area was affected by flood and the total damage was estimated to be 386.59 crores of rupees. (Basic statistics of the NER-1989, NEC). Agricultural lands located in the low and flat alluvial plains, where malfunctions of water retention and retardation are caused by rapid progress of urbanisation, have increased the vulnerability to flood damages topographically.

Micro Credit

Agricultural credit is a crucial input to the farmers living in the remote, hilly, backward and tribal areas. It helps in raising the capabilities of small peasants and promoting sustainable agricultural systems. Because of smaller fragment of land, lack of proper collateral guarantee, most of the commercial banks refuse to give loan to the farmers. Most of the farmers depend upon unscrupulous moneylenders for meeting their requirements of harvesting the crops. Even the government scheme like IRDP failed to provide loans to the farmers to improve their agricultural production and other allied agricultural activities. Without improving the micro credit system, it is not possible to improve the agricultural scenario in Assam as most of the farmers belong to the marginalised category.

Marketing

In Assam no proper marketing facilities exists for farmers either in the form of linkages with markets nor there is any substantial improvement in rural marketing structures. Markets for agricultural products in Assam are underdeveloped. Geographically Assam is in a better position to market its products than other North-Eastern States. But for this there is a requirement to have a better transportation system, communication system and marketing infrastructure, and market intelligence system. Proper marketing system can also act as a source of revenue in the form of tax from the daily and weekly markets and check gates. Brokers and buyers have no access to timely market reports about the price of products on daily basis and in this regard the farmers are far away. Farmers are forced to sell their products immediately after the harvest at a lower rate.

Non-Economic Factors

Lack of proper education to the farmers, disturbed law and order situations, political and administrative instabilities, growth of population, lack of skilled manpower, lack of entrepreneurial skill specially among the youth and the vision of the people to invest in agriculture are some of the principal reasons to contribute adversely for the growth of agriculture in Assam. Along with these the potentiality of Assam in agro-food industries still remains unexplored.

Conservation of Land, Water and Biological Resources

There is no optimal use of land, water and genetic resources in a sustainable manner in Assam. Efforts are needed for the use of marginal, wasteland and community lands for agro forestry, which can substantially, improve the economy of the people. The government should initiate pilot projects on dry land for agriculture. The land reform measures are far behind the other States, which is a must for helping the poor by way of distributing land to the landless and also for agricultural growth.

Consumption of Fertilisers

The fertiliser consumption per hectare of gross cropped area in 1994-95 was 9.5 kg in Assam as against 175 kg in Punjab, 136 kg in Tamil Nadu, 126 kg in Haryana and 75.7 kg on an average in the country as a whole. The pesticide consumption was 0.8 per cent while the other NER consumed 1.2 per cent of the country's total pesticide consumption.

The Success Areas

The Shallow Tube Well Irrigation Scheme implemented in the State under *Samridha Krishak Yojana* (SKY) by the State Agricultural Department is a success in terms of its physical achievements—added irrigation, increased farm output, and higher income for farmers besides efficiency in financial utilisation. The scheme was launched in October 1999 with an outlay of Rs. 230 crore meant to cover 18 districts with the installation of one lakh shallow Tube Wells by March 2001. The primary objective of the scheme was to boost agricultural production, especially during the *Rabi* (winter) season and thereby creating additional employment opportunities and promoting income generation in the agricultural sector. The impact of the scheme can be seen in terms of production of rice, which increased from 32.54 lakh MT to 38.61 lakh MT to 39.98 lakh MT to 40 lakh MT, from 1998-2002 showing an increase trend in rice production in successive years. The production could have been higher, but many farmers shifted to multiple cropping patterns as they were assured of regular irrigation. A boost in farm output and increase in job opportunities were the two direct benefits the farmers derived from the scheme. While the technical rationale was to increase irrigation resources to enhance production, the method was popular participation through cost sharing between the State and the beneficiaries on 2 : 1 ratio. Since irrigation facilities in the pre-project situation were grossly inadequate, STWs turned out to be appropriate and cost effective methods of minor irrigation in the region, which has relatively high ground water levels. From November 1999 to January 2001, the physical achievement of the project was 82,666 against a targeted one lakh STWs in less than 14 months, indicating an achievement of 82.67 per cent.

The percentage of area under irrigation from STWs was 3.34 lakh hectares, which was 20.45 per cent of the total gross area (under crops) of 16.32 lakh hectares. And increase in gross production due to irrigation from STWs was 3.35 lakh hectares, which was approximately 9.61 per cent of the production level. The cropping intensity achievement was 144.09 per cent, though short of the project target of 200 per cent for flood affected and 300 per cent for flood free areas. Additional area under irrigation contributed by STWs rose more than 20 per cent and the estimated additional production towards paddy, *Rabi* crops, and all crops was 4.23 lakh MT, 3.81 lakh MT, and 8.03 lakh MT respectively.

Developmental Strategy

Resource Flow

Assam needs special attention in the area of flow of resources and higher level of subsidies for exploiting the tremendous regional potential. The region has the potential to improve in the agricultural sector. The constraint areas lie in lack of skilled manpower, infrastructure development, resources/credit flow and high transportation cost, etc. needs to be addressed to. Extension services through demonstration and training to the tribal growers assume considerable importance. A substantial portion of investment is needed on—Far Water management for increasing production and for increasing consumption of fertilisers. The multi-agency network consisting of co-operatives, commercial banks and regional rural banks should play a dominant role in agricultural credit disbursement.

Watershed Management and Integrated Management of Reservoirs

In the rain fed areas, watershed management is important factor in improving the agricultural production. Integrated farming systems should be practised in Assam through self-help groups in the micro-watersheds to institutionalise people's participation. The agricultural production strategy should focus on flood control, drainage management, improve irrigation facilities particularly minor irrigation and a better input delivery system supported by extension. The fertiliser consumption will have to be substantially steeped up an the gross irrigated area in this region to be increased to cover at least 50 per cent of the GCA of the districts. Along with the chemical fertilisers, the use of organic manures and bio-fertilisers should be also promoted for better soil health. Minor irrigation holds the key in this region and there is considerable ground water potential. Specially drip irrigation and sprinkler irrigation should be promoted in this zone. As already stated Million well schemes can also make useful contribution. The emphasis should be on raising the capability of small peasants.

There is a need to support integrated management of reservoirs and sediments by developing economical and safe technologies and by effective environmental assessment tools to ensure sustainable water resources. This also can help in land-use planning, and water-shed management with farmer's participation of NGO's as facilitators. The farmers also are to be trained in drainage management.

Flood Control Measures

The flood control measures in Assam has to be a joint effort as the catchments area of river Brahmaputra is located at Arunachal Pradesh and the river Barak in south Assam lies in Manipur.

This calls for preparation for integrated flood control measures for controlling the frequency of the floods and development of water resources. At present there is no integrated flood catchments area treatment plan for flood mitigation in the various catchments area including provision for flood cushions. Therefore, the government and other related institutions should adopt well-organised structural and non-structural measures to alleviate flood damages and prepare operational frameworks for the risk management. Hydrosolidarity should be the basis for wise water governance and hence should provide a background for balancing upstream against downstream water use. This requires an understanding of the water circulation through the catchments.

Due to natural and anthropogenic causes, the climatic pattern in future would be at extremes, causing frequent flash floods in some areas and drought in others. Any model on climate will not be able to predict the daily or weekly amount of rainfall over a given area. Under these circumstances, the only key issue is to stabilise and protect the farmer against the vagaries of climate, through watershed management or watershed development. Watershed and micro watershed must be promoted not only for providing employment, minimising soil erosion, controlling the peak hydrograph, but also enhancing ground water recharge from rainfall.

Micro Credit

The Rural Infrastructure Development Fund (RIDF) which was envisaged in the Ninth Plan should be used to promote projects which encourage organisations of small farmers, artisans and landless labourers for skill upgradation, processing of agricultural products, transport infrastructure, and quality improvement in the agricultural products.

Potentialities in Agro Areas

The potentiality of Assam in agro-food industries still remains unexplored. There is a need for infrastructure facilities like cold storage and processing facilities close to the production centres in rural areas. The government should encourage agro processing and agro industries in the region through a suitable industrial policy.

Sustainability in Agricultural Growth

It is clear that there is little scope for further expansion of the net sown area and that land scarcity will become an acute feature of the rural economy in Assam. Water is a precious asset and there are severe concerns regarding proper utilisation of water resources. Therefore, a judicious use of land and water resources will have to a central concern for sustainability of agricultural growth. There has been a growing concern in recent years about the deteriorating conditions of soil health and water resources due to improper management. The deterioration in land and water resources has been in the form of land degradation and waterlogging. There is a greater need to have an integrated approach in the management of agricultural nutrients, chemicals and in taking effective measures to deal with the problems.

Sustainable Development and Agriculture

The concept of sustainable development emphasises on a symbiotic relationship between consumer human race and producer natural resources. The goal of sustainability is for equity and social justice, economic efficiency, ecological harmony and endogenous choices. The strategy of sustainability includes improving the quality of human life while living within the carrying capacity of the supporting ecosystem. Sustainability is compatibility between ecology and economics. Sustainable development must meet the needs of the present generation without compromising the ability of the future generations with reference to their aspirations and needs. Therefore, the exploitation of resources, the direction of investments and institutional changes are to be consistent with future as well as present needs. Sustainable development in agriculture is promotion of ecological sound agriculture with reference to land, water, energy, nutrient supply, genetic diversity, pest management, post-harvest systems and location specific research and development.

The deterioration in the agricultural base in Assam in spatio-temporal terms, and the irrational management of land have damaging repercussions and are reflective of unsustainable policy frame and planning strategies, which are evident from the observable signs as listed:

(a) Extensive deforestation and *Jhum* farming.
(b) Recurring floods and resultant soil erosion.

(c) Growing land degradation reflected through wastelands and waterlogging areas.
(d) Exponential growth of unemployment in rural areas.
(e) Feminisation of agriculture.
(f) Population growth outstripping available resource base of crop lands.
(g) Adverse land labour ratio.
(h) Mono cropping.
(i) Irrigation practices without proper arrangement of drainage, which would further give rise to alkaline or saline soil through waterlogging.

The unsuitable activities in agriculture includes:

1. Intensive cultivation of land without taking adequate care of soil fertility and soil structure.
2. Indiscriminate application of variety of pesticides, fungicides, herbicides, etc.
3. Indefensible and excessive trapping of underground water.
4. Incurable and rapid replacement of the indigenously adopted varieties with high yielding hybrid strains.

Fortunately for Assam, the advantage of foreseeing a green revolution and sustainable agricultural development lies in the absence of the above-mentioned four points. It only requires a holistic approach for achieving ecologically friendly sustainable development by maintaining the fragile balance between productivity functions on one hand and conservation activities and practices by the concerned farmers through monitoring and identification of problem areas on the other. It requires application of alternate agricultural practices, crop rotation, crop diversification, use of appropriate amount of fertilisers, energy efficient farming methods and reclamation of underutilised or wastelands. For achieving such a self-reliant status or sustainable development in agriculture, the wholistic approach on every resource sector needs to be concentrated at all local and regional levels.

Research and Development

For empowerment of farmers and sustainable agriculture the following areas may be looked into:

- Promote organic farming among the interested farmers through proper technology and guidance.
- Create awareness programme on bio-fertilisers, vermin-compost and organic manure to maintain crop production and also maintain soil fertility.
- Establish agricultural technology information centre with video conferencing facility.
- Provide information technology and all kinds of advisory service to farmers through the single window clearance system.
- The extension centres at each district should have laboratories and the scientists should test any plant with disease.
- The farmers should have direct access to the resources at the centre in terms of technology and advice with a provision of a system for feedback.
- Make provision for sales counter for seeds, fertilisers at the resource centre.
- Make provision for audio-visual presentations and video conferencing with agricultural organisations, scientists, or farmers from any part of the State.
- Make provision for input testing laboratories in all the district headquarters.
- Make provision for training-*cum*-demonstration centres on different agricultural products at the resource centres.
- Establish seed farms for cereals, pulses, oilseeds, cotton, etc.
- Demonstrate water-harvesting technology to the farmers and initiate their participation on irrigation management.
- Providing regional facility for training on ecological conservation and scientific use of land and water resources.
- Provide adequate support to agricultural research in the areas of biotechnology, microbiology, genetic improvement of crops including hybrid technology, genetic upgradation and post-harvest technology.

REFERENCES

Algah, Y.K., *Framework Action Plan for Foodgrain Production*, Planning Commission, Government of India, New Delhi, 1988.

Bhalla, G.S. and Tyagi, D.S., *Patterns in Agricultural Development—A District Level Study*, Institute for Studies in Industrial Development, New Delhi, 1989.

Borthakur, D.N., *Agriculture of the North Eastern Region, with Special Rreference to Hill Agriculture*, Beecee Prakashan, Guwahati, July 1992.

Flood Control Measures: United Nations Development Programme, Kyoto, Japan.

Journal of the North Eastern Council (Vol. 13, No. 4, Oct.-Dec. 1993), March 16-23, 2003.

Rao, U.R., "Remote Sensing for Sustainable Development", in *Photovirvachak*, Dehradun, Vol. 19, No. 4, pp. 230, 1991.

Society for Promotion of Waste Land Development, Papers of Workshop on Small Scale Watershed Development, New Delhi, 1988.

The Hindustan Times, September 4, 2003.

The Telegraph, 25th July, 2002.

The Times of India, October 5, 2003.

The Times of India, July 14, 2003.

Turner, R.K. (ed.), *Sustainable Environmental Management*, CBS Publishers and Distributors, New Delhi, 1991.

APPENDIX

Table A.15.1: Classification of Area in Assam

Area (in hectares)/Year	1996-97	1997-98	1998-99
Total Geographical Area	78,43,800	78,43,800	78,43,800
Net Area for Cultivation	24,92,557	24,93,065	25,10,221
Agricultural Land	10,44,644	10,44,644	10,51,381
Barren/Uncultivable Land	14,47,991	14,48,421	14,58,844
Cultivable Waste Land	87,103	85,915	80,194
Net Sown Area	27,44,118	27,51,175	27,01,053
Total Cropped Area	39,88,600	39,93,992	39,40,649
Area Sown More than Once	12,44,482	12,42,817	12,39,596

Source: Directorate of Economics and Statistics, Government of Assam, 2002.

Table A.15.2: Cropping Area

(*in hectares*)

Crop	*Kharif*	*Rabi*	Double crop	Shifting cultivation
Area	29,58,294	17,44,511	14,87,041	4,82,571

Source: Directorate of Economics and Statistics, Government of Assam, 2002.

Table A.15.3: Ground Water Resources and Utilisable Potential in Assam

Ground water resource (MCM) Dynamic	Utilisable ground water resource for irrigation (MCM)	Utilisable ground water resource for drinking & allied (MCM)	Gross draft (MCM)	Balance available (MCM)	State of ground water Dev. (%)
24,719	21,011	3,708	944	20,067	110.96

Source: Directorate of Economics and Statistics, Government of Assam, 2002.

Table A.15.4: Production of Different Crops in Assam

Year/ Tonnes	Rice autumn	Maize	Rice winter	Wheat	OC and millets	Rice summer
(1)	(2)	(3)	(4)	(5)	(6)	(7)
1993-94	5,86,620	11,621	25,55,620	1,00,831	4,534	2,18,904
1994-95	6,19,126	12,640	24,77,196	1,03,632	4,517	2,12,778
1995-96	5,16,031	12,966	26,22,667	95,051	6,002	2,51,324

(*Contd.*)

Table A.15.4: (*Contd.*)

(1)	(2)	(3)	(4)	(5)	(6)	(7)
1996-97	5,20,191	13,039	25,15,751	1,17,073	5,406	2,92,206
1997-98	5,97,478	14,721	24,70,952	1,10,054	5,391	3,14,444
1998-99	5,20,605	13,964	22,88,114	90,509	5,450	4,46,113
1999-2000	5,14,156	14,056	26,92,709	97,583	5,467	6,53,783

Source: Directorate of Economics and Statistics, Government of Assam, 2002.

Table A.15.5: Average Yield of Different Crops in Assam

Year/kg per hectare	Rice autumn	Maize	Winter rice	Wheat	OC & millets	Rice summer
1993-94	937	658	1,443	1,277	500	1,708
1994-95	973	658	1,472	1,290	502	1,615
1995-96	839	681	1,513	1,107	523	1,619
1996-97	853	679	1,467	1,332	520	1,701
1997-98	984	766	1,417	1,299	514	1,797
1998-99	875	705	1,399	1,010	517	1,992
1999-2000	925	713	1,501	1,279	512	2,219

Source: Directorate of Economics and Statistics, Government of Assam, 2002.

Table A.15.6: Area of Different Crops in Assam

Year/ hectare	Rice autumn	Maize	Winter rice	Wheat	OC and millets	Rice summer
1993-94	6,36,535	17,665	17,97,940	78,937	9,062	1,28,187
1994-95	6,46,586	19,194	17,08,195	80,303	9,000	1,31,712
1995-96	6,24,761	19,051	17,59,845	85,837	11,474	1,55,213
1996-97	6,19,595	19,094	17,36,281	87,904	10,394	1,71,796
1997-98	6,07,358	19,216	17,43,321	84,698	10,490	1,74,974
1998-99	5,94,979	19,810	16,35,230	89,591	10,547	2,23,013
1999-2000	5,57,220	19,711	17,93,994	76,310	10,681	2,94,654

Source: Directorate of Economics and Statistics, Government of Assam, 2002.

16

OPENING UP OF AGRICULTURAL SECTOR AND ITS IMPLICATIONS IN ASSAM

— ALOK SEN

India is in its second phase of reform process, started in 1991 to overcome the then foreign exchange crisis. Such structural adjustment policies introduced globally, involves at the national level the apparent "withdrawal of the State from economic activity". The essential aspect of such withdrawal has been active reversal of previous policy packages and implementation of new policy packages under the Fund-Bank (International Monetary Fund and International Bank for Reconstruction and Development) tutelage. In the agrarian sector it requires complete opening up of domestic market, withdrawal of subsidies and protection encouragement to agri-export and commercial agri-business.

The immediate fallout of agricultural reform policy is the huge inflow of cheap food terms of the developed country in the domestic market and the local farmers are forced to survive in the competition without any government support of any form, government remains only a silent spectator. The hard reality in that a country or region which does not enjoy much efficiency in agricultural production and supports of well-built infrastructure (transport, irrigation, power, etc.) can hardly expect to survive in such a situation. Such reality is not unknown to the developed countries which are now pleading for withdrawal of all forms of trade protections in the developing countries. In fact these rich countries have developed their own economies under tariff walls and are still protecting their economies with various measures of tariffs and non-tariff restrictions. The total amount of agricultural subsidy in the O.E.C.D. countries has risen from $ 298 billion in 1988 to $ 326 billion in 2000, which is 31 per cent of the value of agricultural product in O.E.C.D. countries. Subsidy in agriculture in the European union is about 35 per cent of the total value of her agricultural production. In Japan, subsidies in agriculture constitute 59 per cent of the value of agricultural production. The USA is increasing its agricultural subsidy by $ 83 billion over the next 10 years from 2002 (*The Statesman*, September 10, 2003). Together with other measures, such as tariffs and quotas, these subsidies make it difficult for the developing countries to compete in markets of rich countries. Even more damaging, agricultural exports from the rich countries drive small farmers out of business even in their home countries. This threatens domestic food security and undermines export potentials of the poor nations or relatively poor regions. Hence, the complete implementation of agricultural reform will fall heavily on the economy of any backward region. A region like Assam where 75 per cent of her population is dependent on agriculture, and where agriculture itself is in a backward stage, is sure to witness collapse of the basic structure of the economy.

Reform—Some Case Studies of Poor Countries

Since 1980 more than 80 developing countries have been following over various sub-periods a virtually identical package of economic policies under the tutelage of the international lending agencies. They have done so as the obligatory condition imposed on them for borrowing in order to finance their external payments deficits. The net result of such SAP on the economy of borrowing countries as in Latin America is exactly revealed by a paper by R. van der Hoeven (1994). This paper summarises that taking all countries in Latin America which implemented liberalisation and adjustment between 1980 and 1990 the per capita GDP declined by 9 per cent, the minimum wage by 31.7 per cent and agricultural wage by 26.5 per cent. The proportion of the poor rose by 41 to 44 per cent with a larger impact on urban poor. The averages conceal large variations with only a few countries performing reasonably and the rest approaching disaster scenarios with up to 25 per cent decline in per head income (Bolivia, Argentina, Guyana, Haiti, Nicaragua) while countries like Chile, Colombia and Barbados have seen stagnation in per head GDP.

The post-structural adjustment scenario in sub-Saharan Africa becomes clear from IBRD data, which show that during the decade 1980-1989 out of 33 borrowing countries, 12 countries with 40.8 per cent of population of the region registered average annual per head GDP decline ranging from 0 to 2 per cent and another 9 countries 40.4 per cent of total population registered declines exceeding 2 per cent per annum. Thus 81.2 per cent of the total population of sub-Saharan Africa in a total of 42 countries, experienced fall in income per head; taking all 46 countries as a whole per head GDP fell by 1.1 per cent annually over the decade.

In view of the above, this chapter argues that the inherent intention of such sympathy-coated policies of metropolitan centres appears to be restricting domestic income growth of the developing countries and absorption of the exported products of the developed countries. The opening up of markets for metropolitan capital where substitution of domestic productive capacities by imports rather than income expansion appears to be the main mechanism of securing larger markets. The rationale of this objective emerges clearly when we look at the international division of labour in agricultural production, in a historic perspective. The climate-soils specificity of crops and the concentration of natural bio-diversity in tropical and sub-tropical areas, have led historically to a high degree of import dependence by developed countries on imports from these regions and their policy regime *vis-á-vis* third world agriculture has been geared accordingly to sustain their imports-based high living standards. Despite some advances in productivity in the third world and in bio-technology in advanced countries, to-date the sustenance of import-dependent high living standards in the latter appears to have been possible only at the expense of the basic foodgrains production for local third world population.

The inverse relationship between food production and agri-exportable production is a pan-developing countries phenomenon which is particularly marked in those countries which have taken structural loan in the 1980s and have been pushing agri-exportables for a number of years (Patnaik 1996, p. 2429). Probable logic behind this is, Fund-guided contractionary programmes leading to prolonged deceleration in investment in irrigation and agriculture in most tropical developing countries. In the absence of adequate investment tropical land becomes a resource in virtually fixed supply; hence growth in agri-exports can only be at the cost of decline in domestically consumed output. Table 16.1 is a supportive evidence of inverse relationship between agricultural production and agri-exportable productions in some sub-Saharan African countries.

Table 16.1: Annual Growth Rates of Food and Non-food Crops in Selected SAP-implementing Countries in SSA

Country	Period	Non-food	Food
Ethiopia	1986-90	3.7	0.9
Kenya	1975-80	6.5	–9.5
Sudan	1980-85	13.9	–2.3
Nigeria	1986-89	11.3	–6.1

Source: EPW, September 1996, p. 2446.

Here a question of reverse policy becomes more logical: had those adjusting countries pumped large enough investment into irrigation, improved seeds and fertilisers in order to raise grain productivity, then food output could perhaps have been maintained even with agri-exports. It seems that the structural characteristics of colonial syndrome in the agrarian sphere—the growth of exports and decline in the absorption of basic food staple by the poorer majority of local population—is being replicated today all over the developing world as a general trend which is sharply accelerated in the countries following trade liberalisation. Thus it is a very reasonable observation that the regime of primary sector export drive in every third world country undertaking liberalisation and SAP today, is also the regime of mandatory macro-economic concentration, and hence of decline in rates of productive investment. There is thus the remotest possibility of combining high-agri-exportable growth, though of short-span, with high growth of domestically consumed crops and macro level agrarian development in the long-run.

Open Agri Market—India and Assam

Though India has already attained self-sufficiency in foodgrains production, but this only does not guarantee sufficient strength of Indian agriculture justifying the opening up of this sector for the international competitors. It is common knowledge that productivity of Indian agriculture is much less than the agricultural productivity of other major food producing countries. China, which may be our probable competitor in the open agri-market, is placed in a much better position. With only 100 million hectares of agricultural land, China is producing 400 million tonnes of grain while India with its 146 million hectares of agricultural land produces on an average 185 million tonnes of food (*Assam Tribune*, March 10, 1999). If you take the production per hectare of individual crops, it is found that we are much behind other countries. The average production of rice per hectare in India is around 1,756 kilograms compared to 5,470 of North Korea, we are harvesting only 2,117 kilograms of wheat per hectare compared to 7,716 by the Netherlands. Similarly, India produces 1,606 kilograms of corn per hectare compared to 9,091 by Greece. In case of Soyabean and groundnut, it is the same story. The production of Soyabean per hectare India is 804 kilograms compared to 3,450 kilograms in Zimbabwe. As for groundnut, India harvests only 929 kilograms per hectare compared to 4,600 kilograms per hectare harvested by Israel, that is to say that Israel gets 5 times the groundnut per hectare as against India.

Another factor worthy to mention is that India is not using its agricultural inputs to the optimum level. A country like China which has less cultivable land than India has developed one of the best water management system to get the maximum advantage of land. A country like Netherlands which can grow one crop a year because of cold weather and snow, makes the best use of its inputs we get maximum output per hectare.

The lesson which the experiences of other countries in the field of agriculture holds for us is that, it is high time for us to give a serious thought on policy matter before accelerating the SAP further. The alarm has already been rung. The inverse relationship between food production and agri-exportable production has already shown in face like a tip of the iceberg. Table 16.2 illustrates such relationship in India during the five years immediately after the starting of liberalisation process.

During this period, the foodgrain growth has shown a declining signal of 1.71 per cent of annual growth rate and more seriously the total area under agriculture during 1990 to 1995 has declined by an extent of 4 million hectares (*Economic Survey,* 1995-96).

Table 16.2: Selected Items of Agricultural and Marine Output 1991-92 to 1995-96

Items	Average of annual growth rates (%)
Rice	1.94
Wheat	1.52
Pulses	0.90
Coarse Grains	0.27
All foodgrains	1.27
Oilseeds	4.93
Marine products	6.56

Source: Economic Survey, 1995-96.

The above discussion reveals that Indian agriculture still needs to go a long way in order to make it ready to face the challenges of structural adjustment policy (SAP), the untimely adoption of SAP has been reflected in the form of inverse relationship in the growth of food and non-food items in India. Now in order to assess the condition of the agriculture of Assam in the change scenario the chapter throws light on the present status of agriculture of Assam in relation to that of India.

The most striking fact of Assam's economic development is that it is lagging far behind the rest of the country in almost all fronts. The State's economy as reflected by per capita income,which was 4 per cent above the national average in 1951-52, has deteriorated to the extent of 41 per cent below the national average in 1998-99. The major responsibility of such dismal affair falls on the agriculture since this is the major sector of the State's economy on which 69 per cent of the State's working force dependent for earning their livelihood and the State exchequer depends for 32.52 per cent of income (*Economic Survey of Assam*, 2001-2002). But agriculture in Assam exhibits most of the characteristics of under-developed backward agriculture, namely, a high dependence on agriculture for livelihood, widespread practice of traditional farming techniques and correspondingly low usages of modern farm inputs, low levels and low growth in productivity and incomes in the sector, widespread prevalence of subsistence cultivation, poor inadequate agricultural infrastructure, and so on.

Table 16.3: Crop-wise Yields and Growth Rates

Crop	Yield levels (kilograms per hectare)						Exponential growth rates (%)		
	1980-81 to 1984-85		1985-86 to 1989-90		1990-91 to 1994-95		1980-81 to 1994-95		
	Assam	All-India	Assam	All-India	Assam	All-India	Acreage	Yield	Production
Rice	1,071	1,350	1,114	1,584	1,313	1,807	0.8	1.8	2.6
Wheat	1,146	1,770	1,059	2,066	1,267	2,388	–2.8	0.8*	–2.1
Foodgrains	1,037	1,080	1,073	1,231	1,271	1,453	0.6	1.9	2.4
R&M	458	624	475	772	521	874	1.4	0.9*	2.3
Jute	1,478	1,467	1,560	1,696	1,721	1,876	–2.1	1.2	–0.9*
Tea	1,501	1,451	1,614	1,630	1,728	1,764	1.1	1.4	2.5
Sugarcane	43,201	57,257	43,602	61,387	40,084	66,736	–2.8	–0.4*	–3.2
Potato	6,566	14,166	5,922	15,037	7,354	15,900	4.0	1.2	5.2
Arecanuts	980	1,083	1,163	1,167	780	1,141	3.2	–2.6	0.6*

Notes: (a) R & M refers to rapeseed and mustard.
(b) The growth rates are all for Assam. All growth rates are significant at 5 per cent level, except those with an asterisk (*).

Source: (a) Data are from *Area and Production of Principal Crops*, Government of India, and *Fertiliser Statistics*.
(b) Calculations are by Goyari (2000).

In respect of productivity levels of various crops in Assam, which is used in this chapter as index of competitive strength in the open market, it can be seen from Table 16.3 that yields levels of all the major crops in Assam are very low and well below their corresponding national average. More distressing is the fact that the gap in yield levels between Assam and the national average has widened for the crops shown in Table 16.3. Further the all crop index of agricultural production in Assam has gown by less than one per cent per annum (average over 1992-93 to 1998-99), in comparison with an average growth of about 2.7 per cent per annum at the all India level over the same period.

Associated with the above, other sectors are: The transport system of Assam is very insufficient and irregular and connected with the mainland only through a narrow corridor subject to frequent interruption during rainy season. The proportion of surfaced roads in Assam is only 26 per cent against country's figure of 46 per cent. The railway route length of Assam constitutes a small fraction of 3.8 per cent of the total railway route length of India. Power crisis poses as serious obstacle in the way of any development venture and capital attraction in this region. The State of Assam accounts for only a small fraction of 0.16 per cent of the total electricity generation in the country, registering huge shortage of 6.9 megawatt as in 2000-2001 (*Economic Survey of Assam*, 2000-01). The bleak future of agriculture of Assam under the changed circumstances is further substantiated by the fact that in spite of having abundant surface and general water resources, the States lagging far behind the national average as irrigation covers only 39 per cent of irrigable area against the national average of about 80 per cent. Any major change in the above scenario is also unexpectable since Government following SAP is proceeding with macro economic contraction and the private investors are reluctant to invest in North-East India. Hence this chapter finds sufficient grounds to forecast a SAP—implementing SSA countries—for the State of Assam if the present policies of the government goes on accelerating. Some recent changes in the agriculture in Assam have already created doubt in our minds as these changes are alike the end results in the SSA countries mentioned in this chapter as case studies. Worthy to mention is: there has been a gradual decline in the area coverage under Autumn Rice in the State. During 2001-02 too, the same fell to 4.96 lakh hectares from the area of 5.40 lakh hectares in 2000-01. Area under winter rice, the principal *Kharif* crop, also shown as decline from 17.77 lakh hectares in 2000-01 to 17.15 lakh hectares in 2001-02. The total production of foodgrain crops in the State has fallen to 40.28 lakh tonnes during 2001-02 as against 41.73 lakh tonnes in 2000-01. But at the same time, showing the inverse relationship between food crops and cash crops, the negative fallout of SAP, the cash crops have shown regular increase. For example the principal cash crop jute, increased to 6.75 lakh bales in 2001-02 from 6.68 lakh bales in 2000-01, sugarcane recorded an increase to 10.11 lakh tonnes in 2001-02 from 9.88 lakh tonnes in 2000-01. Table 16.4 presents a comparative picture of production of food and non-food items in Assam and India during the period of economic reform 1993-94 to 1998-99.

Table 16.4: Agricultural Production (percentage change over previous year)

Year	Assam			All-India		
	Food	Non-food	All crops	Food	Non-food	All crops
1993-94	2.8	4.9	3.5	4.2	3.7	3.3
1994-95	–1.4	3.4	1.4	4.0	6.5	5.1
1995-96	2.7	1.9	2.0	–6.4	2.2	–2.4
1996-97	–1.3	0.0	0.0	10.3	8.6	9.3
1997-98	0.7	2.5	1.3	–3.1	–10.0	–6.3
1998-99	–4.0	–1.9	–2.6	5.8	9.4	7.3
Average growth rate	–0.1	1.8	0.9	2.4	3.4	2.7

Source: (a) Data are from Economic Survey, Assam 2000-2001, Government of Assam.
(b) Calculations are by author.

Table 16.4 clearly indicates that the growth of production of food crops is in constant decline in both Assam and India whereas the non-food items are showing an increasing trend. The situation is more precarious in Assam, a negative growth rate of –0.1 per cent of food crops during the mentioned period. Such result is in complete conformity with the arguments put forward in this chapter. The comparative strength of the agriculture of Assam being far less than India witnessed a major shock of SAP at the earliest. This chapter concludes with the opinion that neither Assam nor India is yet ready to open its agricultural sector and every signal of economy at every stage need to be analysed critically rather than direct swallow of policy packages exported from the metropolitan centres of the world for serving their own interest.

REFERENCES

Acharya Shankar (1995). *The Economic Consequences of Economic Reforms*, Sir Purushottam Das Thakurdas Memorial Lecture, Indian Institute of Bankers, Mumbai.

Ahluwalia, Montek S. (1994). *Structural Adjustment and Reforms in Developing Countries*, Proceedings of a conference sponsored by G-24, New York.

Annual Financial Budget of Central and Assam Government for various years.

Assam Development Report, Commissioner, P & D Department of Assam & Planning Commission of India.

Assam Tribune (Daily), June 16, 1997; December 1, 1998; October 3, 1999, December 20, 2000; January 4, 2002.

Bardhan, Pranan (1984). *The Political Economy of Development in India* (Oxford).

Government of India, *Agricultural Statistics at a Glance*, for various years.

Government of India, *Economic Survey of India* for various years.

Government of India, *Statistics at a Glance* for various years.

Government of India, *Economic Survey* for various years.

Joshi, Vijoy and I.M.D. Little, *Indian's Economic Reforms: 1991-2001*, Oxford University.

Nayyar, D. and A. Sen, "International Trade and Agricultural Sector in India" in G.S. Bhalla (ed.), *Economic Liberalisation and Indian Agriculture*, ISID, 1994.

Patnaik, P., "International Capital and National Economic Policy", *Economic and Political Weekly*, March 19, 2002.

Patnaik, Utsa, "Export-oriented Agriculture and Food Security in Developing Countries and India", *EPW*, Nov. 1996.

Profile of States, Centre for Monitoring Indian Economy, India, March 1997.

Sen, Abhijit, "Economic Reforms, Employment and Poverty—Trends and Options", *Economic and Political Weekly*, Nov. 1996.

17

CHANGING PATTERN OF AGRICULTURE IN ASSAM WITH SPECIAL REFERENCE TO RICE PRODUCTION SYSTEM

— B.C. BHOWMICK, B.C. BARAH and N. BORTHAKUR

Rice is the most important crop in Assam cultivated from time immemorial. Assam is regarded as one of those places where rice cultivation originated. With the passage of time the rice cultivation in terms technology has undergone tremendous changes. Agriculture is the mainstay of the State economy and rice dominates the cropping pattern which alone occupies about three-fourth of gross cropped area with average productivity of about 2 t/ha. Three important types of rice culture are grown in the State of which winter or *sali* rice is the main rice culture in the State which is important in respect of acreage and volume of production. The autumn or *ahu* rice is the second important crop grown in the State that usually competes with jute for acreage since both these crops require similar type of land situation and their growing time also similar. Summer rice or *boro* is grown in selected areas of the State and is progressively gaining importance owing to its higher yield potential. In low land having enough soil moisture in *Rabi* season is ideal for typical *boro* rice cultivation. Of course its area in the State is very limited on the other hand irrigated *boro* is cultivated as an irrigated crop in low and medium-lowland situation. In areas where irrigation potential created is higher acreage especially under irrigated *boro* has shown an increasing trend (Nagaon, Morigaon, Kamrup and Goalpara). Deep water or *bao* rice another rice culture grown in selected pockets under very low land situations. Of course its area in the State has not exhibited any appreciable change over the years and its productivity is also remained more or less stagnant. Upland and shallow land are the major rice areas in the State (Tables 17.2A and 17.2B).

Table 17.1: Population Dynamics in the State of Assam

Particulars	1971	1991	% change
Population	1,46,25,152	2,24,14,322	53.26
Decadal percentage variation	34.95	53.26	
Male (Number)	77,14,240	1,16,57,989	51.12
Female (Number)	69,10,912	1,07,56,333	55.64
Agricultural workers	22,83,698	35,59,117	55.84
SC/ST percentage (Rural)	18.43	20.22	
Agricultural labour to total main worker	13.51	14.89	
Rural population percentage	90.03	88.9	
Mortality rate per 1000 birth	17.30	8	
Rural population density	172	257	
Land-Man ratio	0.21 : 1	0.18 : 1	
Male literacy	36.68	61.87	
Female literacy	18.63	43.03	

Table 17.2A: Land-Use Dynamics in the State of Assam

Particulars	1971-82	1997-98	% change
Reported area	78,52,000	78,50,005	–0.03
Forest (ha)	19,84,000	19,30,289	–2.71
Fallow per cent (include current fallow)	2.19	2.25	2.74
Net cropped area (ha)	27,06,000	27,51,175	1.67
Net sown area (% of reported area)	34.46	35.05	1.71
Gross cropped area	34,60,000	39,93,992	15.43
Land put on non-agricultural uses (ha)	9,14,000	10,44,644	14.29
Barren/Uncultivable (ha)	15,41,000	14,48,421	–6.01
Pasture and land under tree crops	4,31,000	4,12,605	–4.27
Cultivable waste	1,04,000	85,915	–17.39
Cropping intensity	127.86	145.17	13.54
Net irrigated area (% of NCA)	5.16	5.24	1.55
Gross irrigated area (% of GCA)	4.04	3.64	–9.90

Table 17.2B: Rice Area by Ecosystem in Assam

('000 ha)

Ecosystem	Area
Rainfed Rice	
Dry land	215
Deep water (1 m)	100
Upland	892
Shallow (0-30)	892
Intermediate	0.2
Irrigated Rice	
Wet season	189
Dry season	14
Total Rice Area	2,302

This chapter aims at providing idea regarding rainfed rice system and characteristics of rice production technology and pattern of their changes across agro-climatic zones of the State of Assam and to diagnose appropriate policy options for improving its productivity and income of farm family in Assam.

Methodology

This chapter is based on both primary and secondary data collected from different published and unpublished sources.

The changes in rice production systems in the State is tried to be captured from relevant analysis of secondary data and economies of rice production is analysed using the primary data collected from different size group of farms in two sample districts, viz., Nagaon and Golaghat under CBVZ and UBVZ respectively for two consecutive years (1999-2000 and 2000-2001). Mostly tabular analysis is carried out in the present chapter. Nevertheless, Gini coefficient and Lorentz curve, poverty gap analysis and Simpson diversified index have been attempted to analyse primary data to measure inequality of asset ownership, discrepancy between below poverty level and actual consumption level and degree of crop diversification respectively for the sample farm households.

The formula for the *Gini Coefficient* is given by

$$G = 1 + (1/n) - (2/n_2\mu)\,[1y_1 + 2y_2 + 3y_3 + 4y_4 + ... + ny_n]$$

where, y_1 is the highest income, y_2 is the 2nd highest income, y_3 is the third highest income, ..., y_n is the lowest income, n is the number of households, and μ is the mean income.

The *poverty gap index* is estimated using the following model

$$P = 1 / n \sum_{i=1}^{q} (z - x_i / z)$$

where, n = number of sample households, q = number of people whose income is below the poverty line, z = income at the poverty line.

Simpson's diversification index is computed using the model

$$D = 1 - \Sigma (a_{ij} / A_i)^2$$

where, a_{ij} is the area planted to the jth crop by the ith farmer and A_i is the total cropped area of the ith farmer. The index D ranges from zero for complete specialisation in one crop to the maximum value of unity indicating a high degree of diversification.

Results and Discussion

Agriculture has been the major contributor to State domestic product in Assam. Nevertheless, the relative share of agriculture towards the SDP has shown a declining trend over the years. Against the contribution of 51.33 per cent in 50's its share in the 90's declined to only 39.53 per cent at current price. At constant price (1980-81) the share of agriculture revealed a decline from 44.21 per cent in the 80's to 34.59 per cent in 90's (Tables 17.3, 17.4 and 17.5). The ever increasing population pressure has necessitated increase in foodgrain production either through enhancing productivity and/or bringing more area under cultivation (Table 17.1). Prior to Green Revolution, a static nature of rice cultivation was observed in the State. Low use of technology in terms of working capital per ha, lack of irrigation facilities, erratic distribution of rainfall, crop damages caused by flood and drought, insects and pests are some of the reasons of stagnation in rice production and productivity. However, the rice production scenario in the State has shown sign of improvement after 80's primarily due to adoption of seed fertiliser technology and acreage expansion under the crop.

Poverty is deep rooted in the State. Although the percentage of people below poverty level has marginally declined from 42.60 per cent in 1983-84 to 36.09 per cent in 2000 (NSSO) but compared to national level BPL 26.10 per cent it is still very high.

Table 17.3: Compound Growth Rate of Rural Population and Agricultural GDP in Assam

Period	Rural population	AG-GDP	Per capita GDP
1980-85	1.90	2.71	0.79
1985-90	2.13	2.92	0.78
1990-95	2.03	1.56	–0.46
1995-99	1.84	1.93	0.13

Source: National Account Statistics (various issues).

Table 17.4: Percentage Contribution of Agriculture Towards State Domestic Product (SDP)

	1980-81		1997-98	
	Current price	Constant price	Current price	Constant price
Agriculture	44.21	44.21	39.53	34.59
Forestry	2.29	2.29	2.04	1.07
Fishery	1.93	1.93	1.64	2.10
Total agriculture	48.43	48.43	43.22	37.76

Table 17.5: Poverty Scenario in Assam (% BPL)

Period	Assam	India
1983-84	42.60	45.65
1993-94	45.01	37.27
2000 (NSSO)	36.09	26.10

Source: Expert Group on Poverty Estimate, Planning Commission, 1997, NSSO, June 2000.

The cropping pattern in the State did not show any appreciable change (1970s-1990s) over the years. Rice continues to cover about 75 to 78 per cent area. Oilseed occupied 6.3 per cent of cropped area in 70s, which increased to 10 per cent during 80s but eventually declined to 8 per cent in 90s. Pulses and fibre crops showed a gradual decline after 80s (Tables 17.6 and 17.7).

Table 17.6: Per cent Area under Different Crops in Assam (Cropping pattern) Triennium Average

	1970-72	1980-82	1990-92	1996-98
Foodgrain	78	76	75	75
Rice	72	69	69	68
Wheat	2	3	2	2
Coarse cereals	0	1	1	1
Pulses	3	3	3	3
Non-foodgrain	22	24	25	26
Oilseeds	6	8	9	9
Others	16	16	16	17

Table 17.7: Changes in Overall Cropping Pattern in Assam over Decades

(*in %*)

	1970s	1980s	1990s
Cereals (Rice, wheat, maize)	80.6	79.4	80.6
Rice	77.6	75.6	77.5
Pulses	3.5	4.1	2.9
Fibres	6.1	3.9	3.0
Oilseeds	6.3	10.0	8.1
Others	2.1	3.3	4.4

Source: Extracted from IRRI, IFAD, ICAR, IIRR; Rainfed rice; a source book of best practices and strategies in Eastern India, 2000.

The State of Assam is classified into six broad agro-climatic zones. Rice productivity differs across the zones so also the topographical situations, which are highly diverse. Of the 10 erstwhile districts of the State five districts, viz., Kamrup, Goalpara, Darrang, Nagaon and Sibsagar together account for 70 per cent of rice area and contributed 69 per cent of production. Amongst the zones, Lower Brahmaputra Valley Zone (LBVZ) shared 28 per cent of production of State's production from 35 per cent of rice area followed by Upper Brhamaputra Valley Zone (UBVZ) with 22 per cent of production from 18 per cent area and North Bank Plain Zone (NBPZ) with 18 per cent production from 20 per cent area (Table 17.8). The relative seasonal area share was also prominent in the zone. The share of autumn rice was the highest (50% of its total rice area) in LBVZ followed by summer and winter rice with 34 and 30 per cent area (Tables 17.9, 17.10 and 17.11). The triennium average area of rice and non-rice in Assam (1951-98) is depicted in Figs. 17.1 and 17.2.

Table 17.8: Distribution of Districts in Each Zone and Share of Rice Area and Production

Zones	District	Area share (%)	Production (%)
LBVZ	Goalpara, Kamrup	35	28
CBVZ	Nagaon	13	15
UBVZ	Sibsagar, Dibrugarh	18	22
NBPZ	Darrang, Lakhimpur	20	18
BVZ	Cachar	9	11
Hills	N.C. Hills, Karbi Anglong	5	6

Table 17.9: Relative Importance of Rice Culture by Season in Assam (%)

	1970-80	1980-90	1990-97
% Share of Rice Area			
Winter	71	72	70
Autumn	27	26	25
Summer	2	2	5
Total	100	100	100
% Share of Rice Production			
Winter	79	80	76
Autumn	19	17	17
Summer	2	3	7
Total	100	100	100

Table 17.10A: Decadal Change in HYV Rice Area in Assam

Particulars	80's	90's	% change
Autumn rice	2,64,054	2,60,723	–1.26
Winter rice	6,58,200	8,08,405	22.82
Summer rice	63,726	1,03,935	64.26
Total rice	9,85,530	11,73,063	19.03

Table 17.10B: Area under Rice in Assam by Season (in lakh ha)

	1970-80	1980-90	1990-97	1998-99
Autumn	5.76	6.05	6.30	5.95
Summer	0.36	0.55	1.38	2.24
Winter	15.26	16.95	17.68	16.35
Total	21.38	23.55	25.36	24.54

Table 17.10C: Per cent Share of Rice Area by Season and Zones

	Winter	Summer	Autumn
UBVZ	88.6	0.3	11.1
NBPZ	70.2	2.3	27.4
LBVZ	59.6	5.2	35.1
Hills	86.6	0.4	12.9
CBVZ	56.3	18.9	24.7
BVZ	80.9	6.1	12.9

Table 17.11: Relative Importance of Rice (% State area) by Agro-ecological Zones 1990-1997

Agro-ecological zones	Autumn	Summer	Winter	Total
LBVZ	49.75	33.81	30.14	35.22
NBPZ	21.72	8.53	19.82	19.67
UBVZ	7.84	0.95	22.28	17.53
CBVZ	13.31	46.41	10.82	13.38
BVZ	4.60	9.90	10.28	8.85
HZ	2.79	0.40	6.66	5.36
Assam	100	100	100	100

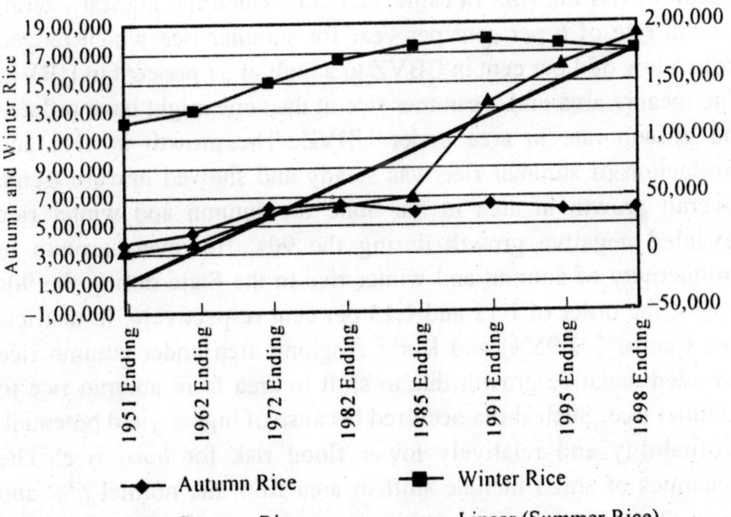

Fig. 17.1: Triennium Average Area of Rice in Assam (1951-1998)

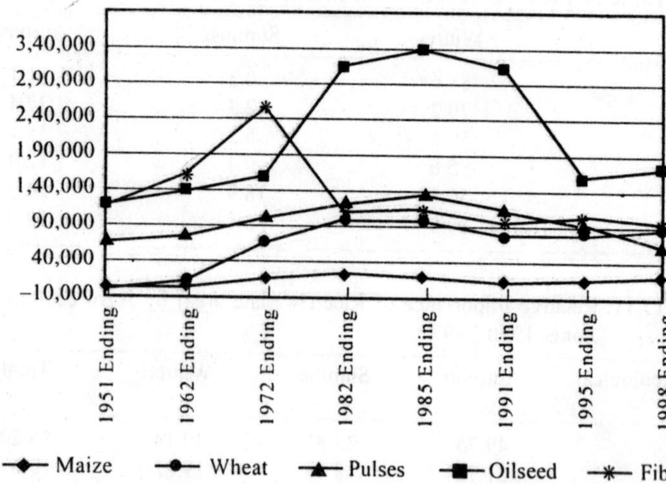

Fig. 17.2: Triennium Average Area of Non-rice Crops in Assam (1951-1998)

Growth and Trend in Rice Cultivation

The decadal growth rates of area, yield and production of rice by season across the zones (Table 17.12) revealed the highest overall growth rate of 6 per cent per year for summer rice which ranged from a low of 4 per cent in CBVZ to a high of 37 per cent in UBVZ. The meagre area under summer rice in the zone might have inflated the growth rate in area under UBVZ. The growth in area and production of summer rice was steady and showed upward trend. Overall growth in area in the State for autumn and winter rice revealed negative growth during the 90s. However, growth in productivity of autumn and winter rice in the State during the 90s was of the order of 1.12 and 1.13 per cent respectively. In districts like Cachar (–2.95%) and Karbi Anglong area under autumn rice revealed negative growth due to shift in area from autumn rice to summer rice. Such shifts occurred because of higher yield potential, profitability and relatively lower flood risk for *boro* rice. The dynamics of shifts include shift in area from the normal *ahu* and early *ahu* to summer rice and jute especially in Kamrup, Goalpara and Nagaon districts.

Table 17.12: Growth Rate of Area, Production and Productivity of Rice in Assam by Season

	1970-80	1980-90	1990-97	1970-97
Area				
Autumn	1.35	–0.06	–0.24	0.57
Summer	2.21	10.86	5.40	5.86
Winter	1.51	0.82	–0.59	1.05
Total	1.47	0.86	–0.17	1.06
Yield				
Autumn	–0.21	–0.03	1.12	0.65
Summer	–2.99	4.01	1.13	0.74
Winter	–0.08	1.16	0.57	1.11
Total	–0.12	1.09	0.77	1.10
Production				
Autumn	1.19	–0.09	0.88	1.24
Summer	–0.85	15.31	6.59	6.64
Winter	1.49	1.99	–0.03	2.19
Total	1.42	1.96	0.60	2.19

Farm Level Analysis

Land Holding

The distribution of sample households of the selected districts (Nagaon and Golaghat) the nature of their holdings, etc. is shown in Tables 17.13, 17.14, 17.15 and 17.16. Over 60 per cent of the farm households in Golaghat and 67 per cent in Nagaon belonged to small and marginal categories with operational holding just over 1 ha in marginal and 1.6 ha in small farms respectively. The difference in size of ownership holdings and operational holdings were substantial and attributed to incidence of tenancy in both the districts. The operational holdings are more equitable than the ownership holdings. The value of Gini coefficient of operational land holdings showed

Table 17.13: Distribution of Farms amongst Different Size Groups

(*in %*)

District	Small and marginal	Medium	Large
Golaghat	60	32	8
Nagaon	67	13	20
All	64	23	14

lesser inequality as compared to ownership holdings. The inequality was more in Golaghat as compared to Nagaon district.

Table 17.14: Average Farm Size by Different Category (ha)

	Marginal	Small	Medium	Large	Total
Golaghat					
No. of farmers	11	32	17	15	75
% of total	15	43	23	20	—
Own land	0.55	1.22	2.41	5.67	1.83
Operational holding	1.03	1.62	2.56	5.67	2.13
Leased in	0.49	0.40	0.15	0	0.30
Tenancy %	47	25	6	0	14
Av. No. of parcels	3	4	6	6	—
Nagaon					
No. of farmers	14	31	24	6	75
% of total	19	41	32	8	—
Own land	0.64	1.33	2.44	5.60	2.34
Operational holding	1.31	1.60	2.64	5.87	2.65
Leased in	0.67	0.27	0.20	0.26	0.31
Tenancy %	51	17	8	4	12
Av. No. of parcels	2	2	3	4	—
All					
No. of farmers	13	32	21	11	75
% of total	17	42	28	14	—
Own land	0.60	1.28	2.43	5.64	2.09
Operational holding	1.17	1.61	2.6	5.77	2.39
Leased in	0.58	0.34	0.18	0.13	0.31
Tenancy %	49	21	7	2	13
Av. No. of parcels	2.5	3	4.5	5	—

Table 17.15: General Household Characteristics in Sample Area

Household characteritics	Nagaon	Golaghat	All
(1)	(2)	(3)	(4)
Number of respondent	75.0	75.0	75.0
Average household size	6.3	6.5	6.4
Average age of household head (years)	54.8	44.5	49.65
Average No. of years in school of household head	8.0	7.8	7.9
Household composition (%)			
Male	58.4	53.5	56.0
Female	41.6	46.5	44.1

(Contd.)

Table 17.15: (*Contd.*)

(1)	(2)	(3)	(4)
Education			
Illiterate	16.9	16.8	16.9
Government training	—	0.6	0.3
Primary	23.6	19.5	21.6
Secondary to High School	42.2	52.1	47.2
College/University	17.2	11.1	14.2
Age group (%)			
< 16 years old	19.7	23.6	21.7
16-50 years old	63.7	62.3	63.0
> 50	16.6	14.1	15.4
Occupation (%)			
Agriculture	46.7	55.2	49.5
Salaried job	8.0	7.7	7.9
Service	6.7	5.2	6.0
Private business	6.7	5.5	6.1
Student	18.9	22.7	20.8
Others	13.1	33.8	23.5
Poverty (%): (Head count; BPL = Rs. 11,800)			
Marginal farmers	66.7	42.8	54.8
Small	2.5	3.7	3.1
Medium	0	9.5	4.8
Large	0	0	0

Table 17.16: Average Area by Farm Size Category and Land Type in Sample Districts (ha)

District	Land type	Marginal	Small	Medium	Large	Total
(1)	(2)	(3)	(4)	(5)	(6)	(7)
Golaghat	Upland bunded	0.32	4.50	8.83	1.20	14.84
	Upland unbunded	0.88	4.39	6.78		12.05
	Medium land	5.83	15.51	21.22	14.79	57.35
	Lowland	7.42	22.96	23.68	18.00	72.06
	Total	14.46	47.35	60.51	33.99	156.31
Nagaon	Upland bunded	—	1.83	5.68	5.35	12.86
	Upland unbunded	0.66	4.48	1.26	12.46	18.86
	Medium land	11.95	35.04	12.74	33.22	92.95
	Lowland	3.54	21.94	14.99	32.80	73.27
	Total	16.15	63.29	34.67	83.83	197.94

(*Contd.*)

Table 17.16: (*Contd.*)

(1)	(2)	(3)	(4)	(5)	(6)	(7)
All	Upland bunded	0.16	3.17	7.26	3.28	13.85
	Upland unbunded	0.77	4.44	4.02	6.23	15.46
	Medium land	8.89	25.28	16.98	24.01	75.15
	Lowland	5.48	22.45	19.34	25.40	72.67
	Total	15.30	55.34	47.60	58.92	177.13
% Rice Area						
Golaghat	Upland bunded	2	9	15	4	9
	Upland unbunded	6	9	11	0	8
	Medium land	40	33	35	44	37
	Lowland	51	48	39	53	46
	Total	100	100	100	100	100
Nagaon	Upland bunded	0	3	16	6	6
	Upland unbunded	4	7	4	15	10
	Medium land	74	55	37	40	47
	Lowland	22	35	43	39	37
	Total	100	100	100	100	100
All	Upland bunded	1	6	15	5	7
	Upland unbunded	5	8	8	7	9
	Medium land	57	44	36	42	42
	Lowland	37	42	41	46	42
	Total	100	100	100	100	100

Cropping Pattern and Rice Production Technology

Rice dominates cropping pattern in sample farms in both the districts. The share of rice area was revealed to be over 80 per cent irrespective of size of farms in both the districts. Pulses and oilseeds emerged as second and third important crop in the cropping pattern of sample farms. Cropping intensity in Nagaon (150%) is comparatively higher than Golaghat (123%) (Table 17.17). Agricultural intensification was possible in Nagaon owing to appropriate policy implication in enhancing the irrigation facilities in summer season (Table 17.18). The varietal diversification in both the districts was rather low. Compared to larger number of traditional varieties modern varieties used by the farmers were fewer in numbers. It has been observed that against 7 modern rice varieties as many as the sample farmers in Golaghat grew 30 traditional varieties and 43 in Nagaon district

(Table 17.19). Amongst the crop establishment methods transplanting is (94%) the main method followed in rice production.

Table 17.17: Cropping Intensity (CI) in Assam by Farm Size Category (% GCA)

District	Marginal	Small	Medium	Large	Total
Golaghat					
Rice area (ha)	13.7	47.2	46.1	28.6	135.6
Net cropped area (ha)	14.46	47.35	60.51	33.99	156.307
Total cropped area (ha)	16.9	59.6	58.4	35.6	170.6
CI	117	126	97	105	109
Nagaon					
Rice area (ha)	15.8	57.2	41.8	85.2	200.1
Net cropped area (ha)	16.15	63.29	34.67	83.83	197.94
Total cropped area (ha)	20.9	69.1	51.8	99.8	241.5
CI	129	109	149	119	122
All					
Rice area (ha)	14.75	52.20	43.95	56.90	167.85
Net cropped area (ha)	15.31	55.32	47.59	58.91	177.12
Total cropped area (ha)	18.90	64.35	55.10	67.70	206.05
CI	123	118	123	112	115

Table 17.18: Irrigated Area under Different Sources in Sample Districts

	Percentage of area under			
	Canal	Deep tubewell	Shallow tubewell	Rainfed
Nagaon				
Marginal	10	—	35	55
Small	2	—	39	58
Medium	0.5	—	21	78
Large	—	—	27	72
Golaghat				
Marginal	—	—	6	94
Small	—	—	23	77
Medium	—	—	16	84
Large	—	—	21	79
All				
Marginal	5	—	21	75
Small	1	—	31	68
Medium	0.3	—	19	81
Large	—	—	24	78

Table 17.19: Adoption of Modern Variety of Rice by Season and Farm Size

District \ Season	Marginal	Small	Medium	Large	All
Golaghat	Measured by % farmers adopting modern variety				
Pre *Kharif*	43	55	29	67	45
Kharif	71	100	92	100	92
Rabi	7	23	21	33	20
	Measured by % area under modern variety				
Pre *Kharif*	78	84	77	82	81
	(1.7)	(5.5)	(3.5)	(2.5)	(13.2)
Kharif	46	68	61	51	60
	(4.3)	(24.8)	(22.1)	(10.9)	(62.1)
Rabi	23	54	70	16	45
	(0.5)	(2.1)	(3.6)	(0.7)	(6.9)
Total	47	69	63	49	61
	(6.5)	(32.5)	(29.2)	(14.1)	(82.2)
Nagaon	Measured by % farmers adopting modern variety				
Pre *Kharif*	45	31	12	20	27
Kharif	73	88	94	100	89
Rabi	27	53	71	47	52
	Measured by % area under modern variety				
Pre *Kharif*	54	45	41	26	41
	(1.8)	(3.6)	(1.6)	(1.2)	(8.2)
Kharif	43	62	62	67	64
	(3.3)	(24.2)	(19.3)	(48.7)	(95.5)
Rabi	21	78	71	55	61
	(1.0)	(8.0)	(5.0)	(4.3)	(18.4)
Total	39	63	62	64	61
	(6.2)	(35.8)	(25.9)	(54.3)	(122.1)

Note: Figures in the parentheses are area under modern variety in ha.

Amongst the sample districts, the weighted rice yield was higher in Nagaon than that of Golaghat being just over 3.5 tonnes in Golaghat compared to 4.5 tonnes in Nagaon (Table 17.20). In respect of varieties more than half of the small farmers continued to grow traditional varieties particularly in marginal size category of farms. In Nagaon percentage of area put under modern varieties was 39 per cent in marginal farms, 63 per cent in small farms and 62 and 64 per cent in medium and large farms respectively. In Golaghat the marginal and small farms put more area under modern varieties (45 and 69%) compared to medium and large farms. The yield of MV was almost at par with that of TV in several cases (Table 17.21). Thus given better care and improved management, the traditional varieties can also perform well. The difference between their yields was small in Nagaon compared to Golaghat. The yield variability across farm sizes and soil type is also substantial in both the districts (Table 17.22).

Table 17.20: Weighted Rice Yield (t/ha)

Year	District	MV	TV
1999	Golaghat	3.4	2.1
	Nagaon	4.4	3.9
	All	3.9	3.0
2000	Golaghat	3.7	2.7
	Nagaon	4.6	3.9
	All	4.2	3.2
Average	Golaghat	3.6	2.4
	Nagaon	4.5	3.9
	All	4.1	3.1

Table 17.21: Yield Differences in MV and TV (t/ha) of Rice

Year	District	Category	MV	TV	All	Difference
1999	Golaghat	Marginal	3.0	1.9	2.4	1.1
		Small	3.5	2.4	3.0	1.1
		Medium	3.5	1.9	2.7	1.6
		Large	2.8	2.1	2.4	0.7
		Total	3.4	2.1	2.7	1.3
1999	Nagaon	Marginal	4.6	4.9	4.8	–0.3
		Small	4.7	4.7	4.7	0.0
		Medium	4.1	2.7	3.2	1.4
		Large	4.4	3.7	4.0	0.7
		Total	4.5	4.1	4.2	0.4
2000	Golaghat	Marginal	3.4	2.4	2.8	1.0
		Small	4.0	3.3	3.7	0.7
		Medium	3.7	2.2	3.1	1.5
		Large	3.3	2.6	3.0	0.7
		Total	3.8	2.7	3.3	1.1
2000	Nagaon	Marginal	4.4	4.5	4.5	–0.1
		Small	4.8	4.0	4.5	0.8
		Medium	4.5	3.7	4.1	0.8
		Large	4.5	3.8	4.2	0.7
		Total	4.6	3.9	4.3	0.7

Table 17.22: Yield of Rice by Soil Type (t/ha)

Districts	Clay	Clay loam	Sandy clay loam	Sandy loam	Loamy sand	Loam	Sandy
Nagaon							
Marginal	4.47 (23.06)	—	2.45 (6.45)	3.80 (70.51)	—	—	—
Small	5.08 (27.25)	4.49 12.72)	4.32 (4.47)	3.88 (50.11)	4.45 (2.92)	4.45 (1.31)	3.11 (1.23)
Medium	4.40 (42.53)	4.51 (11.23)	3.94 (23.72)	3.49 (9.11)	3.02 (7.14)	—	—
Large	4.27 (49.25)	4.51 (9.83)	4.44 (8.75)	3.39 (25.72)	4.71 (0.44)	—	2.64 (2.52)
Golaghat							
Marginal	—	2.56 (29.18)	2.80 (35.51)	2.86 (20.04)	4.04 (6.98)	—	3.53 (2.25)
Small	2.48 (8.57)	2.80 (35.30)	3.25 (11.98)	3.02 (34.80)	2.92 (4.45)	2.99 (3.04)	3.75 (2.35)
Medium	2.70 (5.05)	2.95 (37.73)	2.95 (13.73)	3.05 (39.86)	—	2.86 (1.61)	2.56 (1.40)
Large	2.50 (4.10)	2.83 (18.15)	1.89 (2.97)	3.44 (47.80)	3.59 (25.45)	—	—
All Districts							
Marginal	2.24 (12.76)	1.28 (14.59)	2.63 (20.98)	3.33 (45.28)	2.02 (3.49)	—	1.77 (1.13)
Small	3.78 (17.91)	3.65 (24.01)	3.79 (8.23)	3.45 (42.46)	3.69 (3.69)	3.72 (2.18)	3.43 (1.79)
Medium	3.55 (23.79)	3.73 (24.48)	3.45 (18.73)	3.27 (24.49)	3.02 (3.77)	1.43 (0.81)	1.28 (1.20)
Large	3.39 (26.68)	3.67 (13.99)	3.17 (5.86)	3.42 (36.76)	4.15 (12.95)	—	1.32 (2.03)

Note: Figures on the parentheses are per cent area under each soil type.

Crop Diversification

Crop diversification across farm sizes and districts shows that in *Rabi* and summer season the degree of crop diversification is much higher as compared to main *Sali* season. The main *Kharif* season (*Sali* season) is more or less exhibited complete specialisation since there are not many alternative crop options available to farmers in this season mainly due to land type and apprehension of crop submergence due to flood (Table 17.23).

Table 17.23: Index of Crop Diversification

District/Farm size	*Kharif*	*Rabi*	Pre-*kharif*	Total
Nagaon				
Marginal	0	0.55	0.60	0.50
Small	0	0.60	0.60	0.50
Medium	0.02	0.53	0.68	0.44
Large	0	0.56	0.63	0.37
Golaghat				
Marginal	0	0.50	0.14	0.32
Small	0	0.46	0.57	0.47
Medium	0	0.30	0.48	0.45
Large	0	0.49	0.46	0.54
All Districts				
Marginal	0	0.45	0.56	0.44
Small	0	0.62	0.61	0.49
Medium	0.08	0.54	0.55	0.45
Large	0	0.59	0.63	0.41

Input Use and Return

Labour is the major component of costs in rice cultivation. The extent of labour use in rice cultivation is next to sugarcane and jute (Table 17.24). Gender participation varied from crop to crop and operation to operation. Operations like transplanting, weeding, harvesting, and threshing women labour is mostly used. Use of chemical fertiliser in rice cultivation is rather low in the State. The use of fertiliser was recorded at 25 kg/ha of NPK for traditional varieties as compared to 68 kg/ha for modern varieties against 14 kg and 67 kg in Golaghat respectively (Table 17.25).

The profit analysis carried out for plot specific variables revealed age and education of household head, labour and farm size as important factors in Nagaon. In Golaghat education, land type and soil type emerged as important factors. Amongst the paid out costs cost of seed was substantial and found to be more for MV compared to tall varieties. Fertiliser and manures is the next important factor and its use is three times more in MV compared to TVs. Intensity of input use and the rice yield was comparatively higher in Nagaon than Golaghat. Paid out cost for labour was estimated at Rs. 2,395 per ha against Rs. 1,678 for material cost. The imputed value of family labour was Rs. 4,062 ha against Rs. 2,948 for owned inputs.

Table 17.24: Summary of Cost and Return of Rice Cultivation by Variety

(*Rs./ha*)

	Material inputs	Labour cost	Imputed labour cost*	Total costs	Gross revenue	Net revenue	Return over paid out cost	Yield
Nagaon: Modern varieties								
Marginal	3,598	2,316	1,534	7,472	14,823	7,351	9,443	3.29
Small	2,136	2,446	1,005	5,714	21,517	15,803	17,940	4.78
Medium	2,076	2,344	797	5,461	18,123	12,661	14,500	4.03
Large	1,124	2,166	852	5,095	21,485	16,391	19,047	4.77
Traditional varieties								
Marginal	3,967	5,027	2,913	8,995	16,304	10,223	10,223	3.62
Small	3,961	3,418	1,565	7,379	19,025	13,211	13,211	4.23
Medium	5,492	2,570	1,513	8,062	14,580	8,031	8,031	3.24
Large	1,131	1,706	839	2,838	17,655	15,656	15,656	3.92
Golaghat: Modern varieties								
Marginal	3,791	3,381	3,017	7,171	13,432	6,260	9,278	2.98
Small	4,099	3,256	2,531	7,355	14,847	7,492	10,023	3.30
Medium	5,032	3,409	2,787	8,442	16,490	8,048	10,835	3.66
Large	5,250	3,321	2,646	8,571	17,742	9,171	11,817	3.94
Traditional varieties								
Marginal	3,958	3,245	2,797	7,203	9,974	2,772	5,568	2.22
Small	4,210	2,869	2,303	7,097	10,455	3,376	5,679	2.32
Medium	4,407	3,197	2,580	7,604	11,899	4,296	6,876	2.64
Large	2,980	2,624	2,043	5,603	11,290	5,687	7,730	2.51

* Imputed cost is derived at market price.

Table 17.25: Fertiliser Use Pattern by Different Rice Variety (kg of NPK/ha)

		Nagaon	Golaghat	All
Modern variety	N	42.38	33.09	37.74
	P	9.72	18.71	14.22
	K	16.67	15.55	16.11
Traditional variety	N	15.25	6.25	10.75
	P	1.02	4.67	2.85
	K	9.68	3.55	6.62
Manure (t/ha)	MV	3.53	2.59	3.06
	TV	5.56	2.03	3.80

Income Distribution

Amongst the activities that contributed towards farm income, the share of rice alone is the highest in both the districts. Average annual per capita income in Nagaon district was found to be (Rs. 12,008) higher as compared to (Rs. 11,060) Golaghat district (Table 17.26) income alone contributed 59 per cent of total income of which rice alone constitutes 62 per cent of on-farm income which was 39 per cent of total income in Nagaon. Income distribution by farm size revealed around 41 per cent of the farm households were in the bottom 40 per cent income level and 26 per cent was in the middle 40 per cent income level irrespective of farm size categories. Similar pattern was observed in both the districts. The decomposition of income inequalities by source of income revealed that the share of on-farm income was more as compared to non-farm income in both the districts (Table 17.27). In Nagaon the share of rice was the highest (0.55) followed by non-farm income (0.27) while in Golaghat share of non-farm income was 0.45 followed by rice income 0.34. In Nagaon on-farm income contributed approximately 60 per cent of inequality and non-farm income rest 40 per cent compared to 64 per cent non-farm income in Golaghat. Incidence of poverty as measured by head counts ratio and poverty gap ratio revealed highest incidence in the case of marginal farms 0.67 and 0.05 and 0.43 and 0.13 for Nagaon and Golaghat district respectively.

Changes in Landholdings, Tenancy, Non-farm Employment and Level of Living

Amongst the assets housing alone constituted the bulk of the value

Table 17.26: Average Annual Income by Sources (Detailed)

(Rs./household)

Sources of income	Nagaon	Golaghat	All districts
On farm			
Rice	29,159	21,870	25,515
	(62.21)	(53.75)	(58.28)
Non-rice	15,518	16,500	16,009
	(33.11)	(40.55)	(36.57)
Livestock	2,194	2,322	2,258
	(4.68)	(5.71)	(5.16)
Sub-total	46,871	40,692	43,782
	(61.79)	(56.49)	(59.35)
Off-farm			
Hired labour	440	192	316
	(0.58)	(0.27)	(0.43)
Non-farm			
Remittances	13,973	4,267	9,100
	(48.96)	(13.70)	(30.51)
Forest products	—	187	94
		(0.60)	(0.32)
Handicraft	—	111	55
		(0.36)	(0.18)
Salary	12,186	25,032	18,609
	(42.70)	(80.34)	(62.39)
Others	2,380	1,558	1,969
	(0.98)	(5.00)	(6.60)
Sub-total	28,539	31,155	29,827
	(37.63)	(43.24)	(40.35)
Total income	75,850	72,039	73,925
Average income/capita	12,008	11,060	11,519

of assets. The value of house per household was found to be almost double in Nagaon as compared to Golaghat district. In regard to changes in landholdings approximately 13 per cent respondents reported changes that have occurred due to cultivation being carried out on leased in land 11.33 per cent reported changes due to purchase of land and 10 per cent due to leased out land. The reason of change in tenancy status was reported by 9 per cent respondents due to

Table 17.27: Decomposition of Income Inequality and Marginal Effects by Income Source

Income source	Income share (S)	Gini of source (G)	Relative income inequality (R)	Component contribution (RGS/G)	Relative marginal effect (5) – (2)
(1)	(2)	(3)	(4)	(5)	(6)
Golaghat					
Rice	0.4121	0.3104	0.6587	0.3078	–0.1043
Non-rice crop	0.0787	0.5858	0.2301	0.0388	–0.0399
Off farm	0.1028	0.2488	0.3279	0.0306	–0.0722
Livestock	0.1041	0.4923	0.1295	0.0242	–0.0798
Salary	0.1765	1.3657	0.5096	0.4488	0.2723
Non-farm	0.0793	0.8337	0.4678	0.1129	0.0337
Homestead	0.0466	0.6907	0.3135	0.0368	–0.0097
Total income	1	0.2737	1	1	0
Nagaon					
Rice	0.617	0.337	0.763	0.490	–0.127
Non-rice crop	0.177	0.596	0.490	0.159	–0.017
Livestock	0.031	0.831	0.273	0.021	–0.009
Homestead	0.033	0.848	0.270	0.023	–0.010
Non-farm	0.142	0.892	0.781	0.306	0.163
Total income	1	0.324	1	1	0

break-up of family 7 per cent due to sale of land and 3 per cent due to purchase of lands. The changes in respect of level of living, health and sanitation, school enrolment of children were reported to have improved considerably in the sample households. In respect of food availability 71 per cent and 83 per cent farmers in Nagaon and Golaghat respectively perceived to have improved its availability 75 per cent opined on improvement of health and sanitation, 77 and 89 per cent on overall income level and 76 and 80 per cent on overall welfare (Tables 17.28 and 17.29). The perceived opinion in respect of changes in varieties, crops planted and change in rice yield across farm sizes and sample districts revealed wide variation in rice production in the sample farm households.

Table 17.28: Changes in the Levels of Living (% of responses)

Factors	Nagaon			Golaghat			All Districts		
	Increased	Decreased	Constant	Increased	Decreased	Constant	Increased	Decreased	Constant
Food availability	70.67	8.00	21.33	83.49	3.02	13.49	77.08	5.51	17.41
School enrolment of children	50.67	2.67	46.66	53.52	4.05	42.43	52.10	3.36	44.45
Health, sanitation	74.67	1.33	24.00	74.57	7.15	18.18	74.62	4.24	21.09
Employment outside the farm sector	21.33	9.33	69.33	16.24	7.13	76.63	18.79	8.23	72.98
Employment within the farm sector	34.67	9.33	56.00	31.39	15.26	53.36	33.03	12.30	54.68
Production of food crops	66.67	8.00	25.33	91.89	3.57	4.54	79.28	5.79	14.94
Production of cash crops	24.00	42.67	33.33	73.61	6.53	19.86	48.81	24.60	26.00
Production of livestock	36.00	22.67	41.33	15.78	10.36	73.87	25.89	16.52	57.60
Production of home garden	48.00	14.67	37.33	9.90	14.06	76.09	42.67	14.37	56.71
Overall income	77.33	4.00	18.67	89.55	1.79	8.67	43.44	2.90	13.67
Overall welfare	76.00	1.33	22.67	80.19	—	19.82	78.10	0.67	21.25

Table 17.29: Changes in Rice Yield (% of responses)

Farm size/Season	Nagaon			Golaghat			All Districts		
	Increased	Decreased	Constant	Increased	Decreased	Constant	Increased	Decreased	Constant
Marginal									
Pre-*kharif*	66.67	—	16.67	28.57	—	28.57	47.62	—	22.62
Kharif	16.67	50.00	33.33	57.14	14.29	28.57	36.91	32.15	30.95
Summer	33.33	16.67	33.33	14.29	14.29	—	23.81	15.48	16.67
Small									
Pre-*kharif*	40.00	7.50	25.00	32.14	—	7.14	36.07	3:75	12.50
Kharif	70.00	5.00	25.00	71.43	7.14	21.43	70.72	6.07	23.22
Summer	47.50	2.50	22.50	14.29	—	3.57	30.90	1.25	13.04
Medium									
Pre-*kharif*	33.76	15.38	23.07	33.33	3.03	3.03	32.05	18.41	13.05
Kharif	53.85	23.08	23.08	75.76	15.15	9.09	64.81	19.12	16.09
Summer	46.15	23.08	7.70	18.18	6.00	—	32.17	14.54	3.00
Large									
Pre-*kharif*	43.75	6.25	—	66.67	—	—	55.21	3.13	—
Kharif	87.50	6.25	—	66.67	33.33	—	77.09	19.79	—
Summer	43.75	—	18.75	33.33	16.67	—	38.54	8.34	9.38
All farms									
Pre-*kharif*	45.30	7.28	16.19	40.18	0.76	9.69	42.74	4.02	12.94
Kharif	57.01	21.08	20.35	67.75	11.53	14.77	62.38	16.31	17.56
Summer	42.68	10.56	20.57	20.02	9.24	0.89	31.35	9.90	10.73

Production Constraints

Flood and drought are the two major constraints to agricultural production in Assam. It is interesting to note that despite high rainfall, the drought is a problem to agriculture. Over 35 per cent respondents rank it as the first production constraint followed by flood (47%) in Golaghat. Diseases and non-availability of seed become third and fourth constraints in order of importance. In Nagaon flood was the first severe constraint followed by disease. Besides this, access to information and resources (credit, technology, information and market) are other crucial constraints.

Reasons of Low Rice Productivity in Assam

- Delay and failure in timely supply of improved seeds.
- Climate favouring pest infestations and diseases affecting large areas.
- Scanty use of pesticides.
- High but erratic rainfall.
- Lack of assured irrigation.
- Rainfed farming in over 90 per cent rice area.
- Energy use in agriculture is very meagre (0.4 hp/ha).
- Lack of stress prone modern varieties especially for upland and very lowlands.
- Poor extension machinery, delay and slow process of dissemination of technology.
- Risk of crop loss due to flood/drought.
- Low use of fertiliser.
- Lack of capital investment.
- Inadequate use of inputs in tenant operated farms.
- High input price and non-remunerative product prices.

Policy Options in Rice Production System

Land Related Issues

Land is the most limiting resource for cultivation of rice and per capita availability of land has shrinked substantially over the years. Unfortunately, the information on the distribution of land according to land type is not available either at the district or in the State level.

Substantial area in different seasons is kept fallow in the sample districts as well as in other districts of the State. Such fallowing in most cases is not intentional. Host of social and natural factors compelled the rice growers on land fallowing for instance lack of irrigation facilities debarred farmers from using the uplands in the pre-*Kharif* season while the very low lying areas are kept fallow in the *Kharif* season for fear of submergence of *sali* rice. Stray cattle are serious problem in *Rabi* season for growing *boro* rice since all the parcels are not used for *boro* rice cultivation thus. This factor alone discouraged many farmers in expanding area under or even growing *boro* rice in many pockets. The village panchyats, FMC and NGOs can play important role in addressing these problems. The State department of agriculture, the main agency in the State can also play important role in this sphere.

Rice Production Technology

Seed

Seed is the vital input, which can contribute substantially towards productivity of crop given the other ancillary factors like fertiliser, irrigation pesticides and management. The level of yield obtained in the field is far lower than the level obtained in research fields mainly because of absence of other essential inputs along with variety. Effort has to be taken to narrow down the gaps if not completely bridged through appropriate technological interventions. State's production of seed is inadequate to meet the total demand of seed. The Seed Corporation and State Department of Agriculture can act in appropriate manner to address the problem.

Fertiliser

The average use of fertiliser in the State is only 15 kg against the national average of 77 kg/ha. The use of fertiliser in vegetable and other cash crops are much higher than that of the amount use per unit in rice. The market price of rice hardly attracts the farmers in going for the use of full doses in most cases especially for those farmers who are producing mainly for family consumption. On the other hand increased prices of fertiliser discouraged the farmers who are interested to go for using modern varieties but without fertiliser or using only a partial dose of fertiliser. This problem may be duly addressed by State Department of Agriculture, FCI, FMC and State Agricultural University.

Capital Investment

For development of agriculture a fair degree of capital investment is necessary. However, in agriculture in Assam either for land development, infrastructure or acquiring assets is very meagre especially from the standpoint of farmers' owing to their weak economic background. This is true for smaller size of farms. Most of the infrastructure development that has taken place in the State in different areas are mainly due to efforts of the government. Larger size groups could invest for development of infrastructure and land but hardly do so. Thus agriculture in the State is severely capital starved. This is one of the reasons of backwardness of agriculture in Assam compared to States like Punjab, Haryana, Maharashtra, Andhra Pradesh and Uttar Pradesh. In the case of absentee land owners the *Adhiary Act* enacted by the State Government discouraged them from making any investment in their lands owing to fear of losing ownership rights to tenants. There is tension of losing the right to cultivate the land amongst the tenants and at the same time fear in the minds of absentee landowners of losing of land ownership. The act needs to be modified that will be acceptable to both tenants and absentee landowners. This will help capital investment and crop productivity in the State.

Credit

The credit use in agriculture in the State as on 1998 stood at Rs. 791 per ha against the national average of Rs. 3,633. The figure is indicative of the prevailing traditional nature of agriculture in the State. The farm level data also reflected a meagre amount of credit used in rice cultivation. The number of bank branches per lakh of population in Assam is much lower than the national average. Besides improper utilisation of credit often results in poor repayment and cause huge outstanding and constrained the financial institutions.

Storage and Marketing

The farmers often are deprived of getting due price of their produce owing to their lack of bargaining power, poor market accessibility and lack of retention capacity. Creation of rural storage will help the farmers to store their produce and market them later when prices are remunerative. State Government, Department of Agriculture and State Marketing Board can play important role in this sphere.

NGO and FMC

The NGOs working in this part of the country can play a very productive role to help the most unorganised section of population in the State, i.e., the farmers in the State, educate and influence them in streamlining their production and marketing system. The Farm Management Committee (FMC) are well recognised body of farmers constituted in every nook and corner of the State and can also play effective role in agricultural production and development of agriculture in the State.

Conclusion

Agriculture is the most important sector of the State economy of Assam. Its economy is primarily rural based and about 60-70 per cent of total cultivators are small and marginal farms. The State passed through a period of stagnation owing to low productivity and stunted growth but in the last one to two decades positive changes have taken place. Rice is the dominant crop that covers over 67 per cent of gross cropped area and 84 per cent of net crop area in the State. Rice thus dominates the cropping pattern in Assam. The growth of area under rice (1990-97) was noted to be –0.17 per cent against the productivity and production growth of 0.77 per cent and 0.66 per cent respectively. The rice yield in the farmers' field was noted to be 3-4 tonnes per hectare. The higher productivity of *boro* rice might have inflated the farm level rice yield. Rice is the major source of farm income, which alone contributes over 60 per cent of farm income and over 40 per cent of total income of the farm. Since some of the zones and districts of the State possess larger percentage of area but yield comparatively less as compared to zones and districts having smaller percentage of area under rice but yield better. This would require identification of the constraints that lead to poor productivity. Emphasis has to be given in improving the productivity in the traditionally important zones to improve overall rice production in the State. Changes as perceived by the farmers, in adoption of new and modern varieties and other practices are some of the positive signs of improvement that have taken place in rural farm households. The scenario of rice production in the State is likely to be bright if policy makers opt for appropriate policy in streamlining input supplies and help the farmer growers in obtaining right price for their products.

REFERENCES

Anonymous (2000). *Rainfed Rice: A Source Book of Best Practices and Strategies in Eastern India (Extracted)*, IRRI, Philippines.

Anonymous (1997). *Expert Group on Poverty Estimate*, Planning Commission, NSSO, June 2000.

Bhowmick, B.C. and Borthakur N. (2002). Final Report on *Socio Economic Dynamics of Changes in Rice Production System in Assam*, NATP (RRPS 2), Department of Agricultural Economics, Assam Agricultural University, Jorhat.

Bhowmick, B.C. Barah, B.C. and Borthakur, N. (2003). *Socio Economic Dynamics of Rice Economy in Assam: A 3-tier Disaggregates Level Analysis* (Report on RRPS2) submitted to PIU, CRIDA, Hyderabad.

Bhowmick, B.C., Pandey, S., Villano, R. and Gogoi, J.K. (2000). *Characterising Risk and Strategies for Managing Risk in Flood Prone Rice Cultivation in Assam* (Ed.), Proceedings published by IRRI, Philippines.

18

FARM SIZE AND LAND PRODUCTIVITY
A Study of the Tea Gardens in Upper Brahmaputra Valley, Assam

— T. CHAND DAIMARI

Introduction

India is the largest producer of tea in the world with 846 million kgs in 2000. It produces about 28 per cent of the total tea of the world. Consequently India is only the fourth largest exporter in the world. Export performance has been declining gradually since 1998. Until 1987-88 tea was the most significant traditional export item from India accounting about 21 per cent and ranked first among the agricultural exports, but it started declining and at present it occupies third position by contributing 9 per cent of the total agricultural exports (Boriah, 2002). Similarly its contribution to the agricultural growth in the State has declined from 6.70 per cent in 1954-55 to 0.07 per cent in 1998-99 (Chand Daimari and Singh, 2002). Therefore, it is necessary to make efforts to raise the productivity level, so that a small piece of land can produce more with better farm and technology. Earlier, efforts have been made to increase the volume of production by extending the existing gardens and opening new small gardens. But soon it is realised, that emphasis has to be laid more on the productivity than on the production (Rai and Saikia, 1990). But the very nature of the relationship which existed between farm size and the land productivity is the main structural question to be dealt with.

The Study Area

The study area Upper Brahmaputra Valley lies between 25°12′-28°0′ N and 89°70′-96°0′ E comprising seven districts (Golaghat, Jorhat, Sibsagar, Dibrugarh, Tinsukia, Lakhimpur and Dhemaji). It covers an area of about 21.67 lakh sq km (35% of the total area of Assam). The valley bears transitional characteristics from climatic point of view due to its specific location within the eastern loop of the Himalaya. It is bounded by Eastern Himalaya on the north-east, Nagaland on the south and Sunitpur, Nowgaon and Karbi-Anglong on the west.

Methodology

Present study is based on primary data of 1998 which had been collected from the tea gardens of different physiographic and climatic conditions. Altogether data of 440 tea gardens of different sizes have been collected. To understand the nature of relationship between the variables considered in the study, correlation formula has been used. After processing the data tables are prepared in such a way which will be able to reflect the relationship between the farm size and the land productivity. A systematic study have been put forward by dividing the valley on the basis of physiographic and climatic characteristics into the following zones: (a) the Foothill, (b) the Central Plain, and (c) the Flood Plain.

Farm Size and Land Productivity

Though there exists a strong relationship between farm size and productivity of land, yet the proposition which attracted considerable attention and has continued to recur in the present discussion is the inverse relations between the two. Even if such relation arises, it does not provide sufficient base to judge the relative potentialities of the different size groups. It is an empirical reality that different farmers produce according to different laws of production and it has been found that these laws are following the pattern of land holding according to size classification (Mishra, 1990). Despite these limitations, the inverse relationship acquires some significance in the present study as it could provide some rationale for arguing that small farm size are superior to large ones on purely physical and socio-economic ground.

The land productivity kg/ha is found to be highest in the Central Plain (2,649 kg/ha). The lowest class category consisting of 76 farms contains highest productivity of land irrespective of all the zones. But Table 18.1 reveals that as the size increases the productivity of land decreases. But the decrease is not proportional and as such there is no significant variation between the farm size and the land productivity.

As far as the area is concerned the Foothill Zone occupies largest area (about 70.78%) but land productivity is as low as 2,153.03 kg/ha. Highest land productivity is recorded in the Central Plain which is about 2,649 kg/ha with lowest percentage of area (about 66.98%) in the valley.

Farm size (701-800 ha) category covers about 64.26 per cent area under tea has the lowest productivity of land, only (709.71 kg/ha) which is not only lowest in the Foothill zone, but also in the entire valley. The highest percentage area under tea, about 85.55 per cent (301-400 ha) in the Foothill zone, about 81.10 per cent (101-200 ha) category in the Flood Plain zone, 75.29 per cent (501-600 ha) categories have very low land productivity, 1,246.44 kg/ha, 2,302.27 kg/ha and 1,985.60 kg/ha respectively, which are much lower than normal average of their respective zones.

It is apparent from the above data, that there is a negative correlation between the agro-ecological zones and their land productivity. However, comparing the calculated values with the tabulated values of corresponding figure it appears that the relation is not very significant in the Foothill Zone and Flood Plain Zone. The negative correlation is comparatively more significant in the Central Plain Zone. This unconventional relationship between the farm size and productivity of land clearly reveals certain serious facts.

A number of economies and diseconomies prevail due to homogeneity and heterogeneity of agro-ecological conditions in different sizes of gardens (Visser, 1999). The small gardens have more homogeneity in regard to their physical features, land conditions and land-use patterns, using technically superior method of production, less socio-economic problems, easily manageable and more intensive application of inputs like labour and capital, but diseconomies of production output. But large size gardens have

Table 18.1: Farm Size and Land Productivity

Agro-ecological Zone	Farm size (ha)	Frequency	Area under tea (%)	Land productivity (kg/ha)	Correlation coefficient (*r*)
Foothill Zone	0-100	13	75.55	6,456.53	
	101-200	16	70.70	2,755.59	
	201-300	9	60.48	1,855.26	
	301-400	3	85.55	1,246.44	
	401-500	6	74.28	1,449.44	
	501-600	12	73.09	1,945.15	
	601-700	6	64.39	2,028.46	
	701-800	3	84.26	790.71	
	801-900	3	60.32	1,896.14	
	901-1000	4	67.59	1,931.86	
	1001-1100	8	62.37	1,330.84	
	Total/Avg	83	70.78	2,153.03	–0.579
Central Plain Zone	0-100	41	71.73	6,697.95	
	101-200	46	73.76	2,863.04	
	201-300	21	62.39	2,342.53	
	301-400	21	69.07	2,225.28	
	401-500	17	63.44	1,708.61	
	501-600	15	75.29	1,985.60	
	601-700	17	65.39	2,252.52	
	701-800	14	64.50	2,003.79	
	801-900	1	61.40	2,295.64	
	901-1000	4	65.09	1,367.34	
	1001-1100	21	64.65	1,396.66	
	Total/Avg.	218	66.98	2,648.99	–0.667
Flood Plain Zone	0-100	22	71.85	7,933.40	
	101-200	25	81.10	2,302.27	
	201-300	16	62.98	2,733.05	
	301-400	19	71.79	2,208.55	
	401-500	17	69.68	1,392.75	
	501-600	4	62.34	1,429.70	
	601-700	7	72.41	3,297.26	
	701-800	7	68.75	1,870.61	
	801-900	8	76.65	1,731.53	
	901-1000	2	63.25	1,794.45	
	1001-1100	12	65.30	1,459.60	
	Total/Avg.	139	69.69	2,604.83	–0.581

more heterogeneity of natural conditions of land but restores better scale of economies. Therefore, the amount of tea produced by an individual garden is directly influenced by its areal sizes. As Heady, 1968 observed, that if the size of economies are very great, other end may be extended only at a very great sacrifice. On the other hand, if the scale economies are small or non-existent, small farms might be used with little sacrifice in attaining a more nearly equal distribution of farm wealth, political stability within agricultural and similar ends.

Interestingly a result emerges from Table 18.1 where the land productivity is the highest in the smallest farm sizes (0-100 ha) regardless of the agro-ecological zones. It also reveals that the land productivity abruptly declines as the garden size increases.

Conclusions

The whole picture in the Upper Brahmaputra Valley is more complex than what is apparent. In the context of backward economy, lacking in intensive modernisation (both in cultivation and manufacturing), low prices, high cost of production, and dominance of highly fragmented small holdings have an immediate impact on the productivity of land. Therefore, it is critical to hold the view that in the Upper Brahmaputra Valley, physical and socio-economic conditions have not allowed any strong relationship to emerge between farm size and land productivity as further investigations are required for other factors which may be playing significant role in establishing the relationship between the two.

REFERENCES

Boriah, G. (2002). *India, World's Largest Consumer*, The Hindu Survey of Indian Agriculture, 2002, New Delhi, p. 125.

Chand Daimari, T. and Singh, S. (2002). *Tea Production in the Economic Development of Assam: Environment Energy and Development*, The National Geographical Society, Department of Geography, B.H.U., Varanasi, India, p. 123.

Heady, E.O. (1968). *Economics of Agricultural Production*, Prentice-Hall Pvt. Ltd., New Delhi, p. 349.

Mishra, S.K. (1990). *On an Alternative Measure of Productive Efficiency in Agriculture Sector*, Hill Area Development, The Geographical Society of North-Eastern Hill Region, India, Shillong, p. 120.

Saikia, H. and Rai, R.K. (1990). *Farm Size and Productivity: A Case Study of Nagaon District, Assam*, Hill Area Development, The Geographical Society of North-Eastern Hill Region, India, Shillong, p. 135.

Visser, S. (1999). *Scale Economy Conditions for Spatial Variation in Farm Size*, Geographical Analysis, Vol. 31, No. 1, pp. 27-44.

19

THE IMPACT OF URBAN GROWTH ON THE LAND-USE PATTERN IN RURAL AREAS OF ASSAM

With Special Reference to Agricultural Holdings

— UPALA BARUA

The present global population is half rural and half urban, but the world's cities are swelling. If this trend continues, by 2025, two-thirds of the world's people will live in urban areas, and 80 per cent of these urban residents will live in developing countries. As urbanisation increases, policy makers in developing countries will be challenged to design ways to feed their cities ideally by relying mostly on their countries' own agriculture sectors. They must also work to prevent undernutrition, cope with changing diets among urban residents, and seek to quell the trend toward obesity.

The developing world is now experiencing urbanisation in the way developed countries did in the past, with urban populations galloping in one or two decades. The only difference is that urban dwellers in the developing world earn as little as $ 200 per person as annual income, compared with more than $ 20,000 in the United States.[1]

India has a tradition of having urban centres since pre-historic times. Archaeological evidences prove this. But today the pace at which India is getting urban perhaps nowhere in the history could be found.

Similar is the case with Assam. Assam is steadily growing more urban during the post-independence period. In 1901 there were only 12 towns in Assam; these were Guwahati, Dibrugarh, Silchar, Nagaon, Jorhat, Dhubri, Tezpur, Karimganj, Barpeta, Sibsagar,

Golaghat, and Goalpara. 50 years later, in 1951 the number of towns doubled. But between 1951 and 1991 there were 263 per cent growth in towns—from 24 to 87 towns in Assam. Similarly, population in towns also increased rapidly. In 1901, total population in town was slightly more than 77 per cent, which increased to 3.5 lakhs in 1951 registering a growth of 354 per cent. Population further increased to 24.70 lakhs in 1991 registering a quantum jump of 605 per cent. In other words, there has been steady growth of urban population from 2.34 per cent of the total population to 11.08 per cent during these 9 decades. This is with reference to census towns. As per Government statistics there are at present 122 towns. Thus the total area under town and total population would be far greater than what has been stated above.

In order to understand the characteristics of towns, it is necessary to understand how the towns are declared. Every State has its own statutes, which lay down the process of declaration of towns. In Assam, towns are declared under three statutes. In fact there are two types of towns: (1) Municipal towns, and (2) Revenue towns. Municipal towns are declared under Assam Municipal Act, 1956 while the revenue towns are declared either under Assam Land and Revenue Regulation or under Assam Land Revenue Reassessment Act, 1936. Basically all towns are at first revenue towns and then these are adopted under Municipal Act. Two basic characteristics for declaration of a town under revenue laws are: (1) majority of its population should be engaged in occupation other than agriculture and (2) agricultural land should be less than that of the other category of land. Apart from these there are also other criteria, which are required to be fulfilled, before declaration of a town. But those criteria are not a matter of discussion here.

Once a town is declared, the gradual annihilation of agricultural land gets started. More and more people start entering in the service sector. More and more lands are required for construction of houses, business units, and trading offices and usually for such purposes the agricultural lands become the target, as these comparatively costs low. This is how agricultural lands start coming to markets in towns. In certain cases land transfer become a hot business for some persons. M.S. Rao writes, with the expansion in the area, the city generally absorbs the rural villages, when the village lands are sold off, at lucrative prices, for meeting the demands of urban development.

The villagers thus devoid of land, become landless with no work to fall back upon, more towards cities in search of employment. . . .[2] As trade grows, migration from the nearer villages in search of jobs and earning cash begins enhancing the population of the area. Demand for residential houses grow—renters finding an opening start business of rented houses. This is the basic character of a growth of urban centre in Assam.

Thus, such unplanned urban growth in the State has been causing irreparable loss to the agricultural lands by way of acquisition for construction of factories dwelling tenements, etc. Moreover, large-scale increase in population has also seriously affects the land holdings of farmers and they are becoming farm labourers in their own lands. Due to lack of any effective scheme to develop the wasteland, there was hardly any effort to compensate the loss generated by growth of urbanisation. The official statistics shows that total land area utilised for crop remained almost same over the last decade. In 1981-82 total cropped area of the State was 3,910 thousand hectares, which remained almost same in the year 1990-91 at 3,962 thousand hectares. While during this period population of the State rose at an average of 2.5 per cent per annum. While talking about conversion of agricultural lands to other uses generally two trends can be observed:

Table 19.1: How Agricultural Land Decreased in Towns

(Area in acres)

Sl. No.	Town	Area of agricultural lands 1978	Area of agricultural lands 1990	Net decrease (–)	Percentage of reduction
1.	Basugaon Town	556.90	243.81	–313.09	56.22
2.	Dudhnoi Town	696.02	575.32	–120.70	17.34
3.	Lakhipur Town	475.00	395.82	–79.18	16.67
4.	Goalpara Town	862.29	678.35	–183.94	21.33
5.	Abhayapur Town	139.02	85.32	–53.70	38.63
6.	Mancachar Town	660.16	433.89	–226.27	34.28
7.	Bongaigaon Town	99.72	35.76	–63.96	64.13
8.	Krisnai Town	232.88	87.53	–145.35	62.41
9.	Gauripur Town	185.31	79.41	–105.90	57.15
10.	Sapatgram Town	319.06	104.01	–215.05	67.40
	Total	4,226.36	2,719.22	–1,507.14	35.66

(1) conversion within the towns, and (2) conversion outside the town as an impact of urbanisation. Below it is shown how the agricultural lands get reduced in 10 towns in the erstwhile Goalpara district of Assam. In terms of urbanisation, the position of Goalpara district is in the lowest category. Till 1991 there were no Class II towns in the erstwhile Goalpara district. Dhubri was a class III town while Goalpara and Bongaigaon were Class IV towns in 1971. Despite that the rate of conversion of agricultural land in towns is significant.

It may be observed from Table 19.1 that in 12 years period as much as 35.66 per cent of the total agricultural land was reduced, i.e., almost at the rate of 3 per cent per year. In no cases there were increments. This gross reduction of agricultural land takes place mainly due to conversion of agricultural lands to other uses. Table 19.2 shows the break-up of converted agricultural lands with reference to their current use.

Table 19.2: Break-up of the Agricultural Area Converted in Towns

(Area in acres)

Sl.No.	Agricultural land converted to	Area	Per cent
1.	Best Residential class	427.23	28.35
2.	Mediocre Residential class	378.52	25.11
3.	Trade site	489.62	32.49
4.	Industrial Site (Brick kiln)	211.77	14.05
	Total	1,507.14	100.00

It may be observed that major part of the converted land was in favour of trade site followed by best residential class. If we consider both the residential classes together, then it becomes the highest class to which major part of the agricultural land was converted (53.46%). As per Government Land Policy announced in 1989, which is still in vogue, no land is allotted or settled within the towns for agricultural purposes. As such, enhancement of agricultural land within the town is an unrealistic proposition. Because the only source of enhancement, i.e., by bringing the government waste land under cultivation has been blocked. While, for reduction of agricultural lands, all roads are open. There is an executive instruction number 6 in the Assam Land and Revenue Manual, which prohibits transfer of agricultural lands for non-agricultural purposes. But this is not valid in case of towns. Even in case of rural areas the agricultural lands can be converted to other uses after obtaining permission from the Deputy Commissioner.

I would like to present here the data I collected during the course of one of my field study in Kamrup and Nagaon district for a project entitled "*Extraction of Alternative Resources by a Tribe under Subsistence Economy: A Case Study*".[3] The finding of the project was astounding. It was a village study and in both the villages I have discovered how people by selling their land first became landless and then resorted to alternative means of livelihood especially beer making and selling. Both the villages were located within two to three km from the nearest towns. This would help us understanding the impact of urbanisation on agricultural lands and how village lands get converted to other uses.

The agricultural land gets reduced to in the nearby villages of an urban locality in two ways: (1) by transfer of land by way of sale, mortgage or acquisition by Government for establishing Public Health Centre, or other Government offices, etc., and (2) by converting the use of land by the land owner.

Table 19.3: Conversion of Agricultural Lands in Villages

(Area in acres)

Village	Total agricultural land in the village (As in 1975)	Converted by way of transfer (As in 1995)	Converted by the landowner (As in 1995)	Total reduction	Per cent
Village-I Kamrup	92.55	29.05	15.52	44.60	48.19
Village-II Nagaon	121.63	16.20	19.43	35.63	29.29

Thus it may be observed that the rate of declination of agricultural land in respect of the village-I is 2.40 per cent annually while in respect of the village-II it is almost 1.5 per cent annually. If this rate continues, all the agricultural lands in village-I will get annihilated within next 22 years. The conversion of agricultural land for other uses has a unique pattern. In most cases it was transformed to residential plots by the persons who migrated to towns and engaged in various types of jobs. This is similar to what can be called the Rao effect—Rao says: the villages who migrate to towns for jobs cannot severe their tie with the village as they leave their families behind to be supported by remittances from the urban areas. These village migrants built fashionable houses in their native villages invest money in land and industry, etc.[4]

As per Government statistics, total land area put to non-agricultural use was 8,81,000 hectares in 1977-78, which increased to 10,51,381 hectares in 1998-99 giving a rise of 19.38. This evidently causes negative impact on agriculture.

Despite all these, very little has been done to protect the agricultural land from being annihilated by urban growth. The above analysis clearly reflects that government intervention is needed to regulate agricultural land. While it is beyond doubt that cities will be the net importers of food and other agricultural raw materials, agriculture possess a challenge not only for rural agriculturists but for city people as well. This is because there is nothing that can prevent urbanisation. As per FAO's[5] projection, Asia will be home of 12 of the 15 largest cities of the world. It is in this context that the concept of urban agriculture is getting wide importance today. In a recently held e-conference on "Optimising agricultural land-use in the city area: Access to land and water, adequate norms and regulations, integration in land-use planning" (3-22 November, 2003) it was resolved that a city with effective use of its available land resources could be able to feed a section of inhabitants especially the urban poor. In this context a pilot project has already been coming up in Chgayan de Oro of Southern Philippines, to establish four pilot allotment gardens in different parts of the city with financial assistance of Europe Aid's Asia Urbs Programme.[6] Those gardens shall serve as a model and learning centre to enable a future extension. I believe, this idea fits well to our situation as well and, is therefore, worth importing. Perhaps it is high time we reorient our conventional attitudes in respect of agriculture.

NOTES

1. From the speech of Wally N'Dow, Secretary-General of Habitat II in the Third Conference.

2. Rao, M.S.A., *Urbanisation and Social Change*, Orient Longmans Ltd., New Delhi, 1970, 2-6.
3. Barua Upala, "Ethnic Groups", in Indira Barua, S. Sengupta, Dutta Das D. (ed.), *Cultural Continuities and Social Change*, Mittal Publications, New Delhi, 2002.
4. Rao, M.S.A., *Ibid.*
5. FAO, 1998b. *Urban and Peri-urban Agriculture: Policy Issues and FAO's Role*, COAG 99/13, Rome, FAO.
6. Robert, J. Holmer, Merlito, T. Clavejo, S. Dongus, Axel Drescher, *Allotment Gardens for Cagayan de Oro City: Improved Access of the Urban Poor to Land for Agriculture.*

20

LAND CEILING ACT IN ASSAM WITH SPECIAL REFERENCE TO AGRARIAN REFORMS

— SHABEENA YASMIN SAIKIA

Assam has a land area of 78,43,800 hectares out of which only 39.40 lakh hectares is the gross cropped area constituting 50.24 per cent of the total area. As per an estimate only 80,000 hectares are lying as cultivable wasteland. Today agriculture provides 32.30 per cent to the Gross State Domestic Product. A fact of great importance about Assam's agriculture is that small and marginal farmers constitute 62 per cent.[1] The majority of the farm has the size ranging from 0.02 and 2.39 hectares. There are only 2,379 numbers of holdings whose farm size is 20 or more hectares.

This is the state of affairs after almost 45 years of land reforms in the State. Today the agricultural farm size is gradually getting even smaller due to (i) demographic growth which has eliminated the scope of having any excess land, (ii) transfer of land holdings, and (iii) competitive use of land due to urbanisation. The State's experience in respect of redistribution of land shows very little progress under existing land reforms programmes. At present several numbers of State sponsored agrarian reforms schemes are under implementation in Assam; prominent among these are (a) implementation of the Assam Fixation of Ceiling on Land Holding Act under which land holders having land in excess of 6.67 hectares are to surrender the same in favour of the landless persons, (b) implementation of Assam (Temporarily Settled Areas) Tenancy Act by which the State wishes to abolish the existence of intermediaries and confer the ownership rights to the tenants who cultivate the land, and (c) consolidation of Land Holding Act under which fragmented land holdings were to be consolidated.

This chapter attempts to analyse the land reforms measures to bring about agrarian reforms, as adopted by Assam Government in the State. The main focus will be on the operation of the land ceiling legislation and its evaluation on the basis of current data. This would be done in a historical perspective as the history of formal adoption of land reforms programme in Assam can be traced back to 1950s.

The Concept of Ceiling

The concept of ceiling on land holdings as a land reforms measure came first in May 1955. The Planning Commission set up a panel on land reforms to make an indepth study of the issue and to suggest measures to be taken up in this connection. The penal after study of the issue unanimously accepted the principle of ceiling on land holdings that "among all resources, the supply of land is the most limited and the claimants for its possession are extremely numerous. It is, therefore, obviously unjust to allow the exploitation of any large surface of land by a single individual unless other overwhelming reasons makes this highly desirable.[2] Evidently the socio-political situation prevailing at that time backed this concept of redistribution of land. Pandit Nehru advocated the redistribution saying that *swaraj* means fundamental changes in the land system, which relieve or lessen the burdens and provide land to landless.[3] The panel also recommended that ceiling should be applied to other sectors of the economy, so that there may be an equitable distribution of wealth. The two main economic aspects of this scheme of redistribution were: (1) to provide a venue of employment of maximum number of people in the rural sector, and also (2) to provide them a means of assured annual income.[4]

In tune with this aspiration expressed at national level, Assam had also enacted the Assam fixation of ceiling on Land Holdings Act in the year 1956, i.e., immediately after the concept was mooted. Assam was in fact a pioneering State in this line. The basic objective of this Act as stated in the preamble of the Act was to impose limits on the amount of land that may be held by a person. Thus, the acquisition of excess land was not the prime object but an objective that follows from the above objective only.

The social aspect of the legislation was to ensure an equitable distribution of the land amongst families dependent on cultivation. It was envisaged in the Act that with the enforcement of its provisions, the difference between big and small farms would be minimised. The Act was brought into force on 15th February, 1958 in the seven plain districts of Assam. Later it was extended to all parts of Assam except that of the Karbi Anglong and NC Hills.

Level of Ceiling

The Act in Section 4 specifies 16.6 acres of land as ceiling limit. Originally, the ceiling limit set was 49.58 acres; but later the Act was amended in 1971 to make it 24.79 acres and again in 1972 to make it 16.6 acres for individual holdings. It is doubtful whether the Government while fixing the ceiling limit had taken into account of the criteria fixed by the Land Reforms Committee. The Panel in its report provided that a family holding should ensure the minimum income necessary for supporting a family of at least five persons. The data on national income available at that time revealed that the annual income per earner engaged in agriculture was Rs. 500. The Penal assumed that there were 2 to 2.5 earners in a family of an agriculturist. Thus a family obtaining an income of Rs. 1,200 per annum for agriculture could be termed as in possession of an economic holding. The Panel was of the view that the land ceiling should be determined in such a way that a family may even thereafter obtain annual income equal to three times the income of an economic holding, i.e., he may obtain a net income of Rs. 3,600 per annum which was sufficient for a reasonable living according to price indices of that period. Considering the above recommendations of the Penal, the ceiling limit for a family according to the price indices of 1955-56 was about 30 acres of fair quality of land.

It is a fact that during 1950-51 the per capita income of Assam was higher than the national average by Rs. 50 only. At the time of enactment of the ceiling Act, the limit for land holding was fixed at 49.58 acres, which was reasonable. Average productivity per acre of land at that time was 28 maund which gave an average income of Rs. 2,500 to 2,800 which was far below the amount fixed by the Penal. Considering the price of the output which was much less than the national average and the cost of input which was higher than the national average the ceiling limit fixed was not reasonable one. But the problem was that very few peasants were actually there who had more than 49 acres of land. As a result the ceiling Act would have been a mere statute in a showcase. Till 1966-67 under the Act total land acquired was only 20,437 acres out of which possession of only 2,837 acres could be taken over by Government[5], i.e., in other words the performance rate was only 13.88 per cent. At that stage 41.7 per cent of the households in Assam were either landless or owned less than 0.05 acres and another 20.02 per cent had holdings below 5 acres.[6]

In 1970, the Act was amended and the ceiling limit was reduced to half, i.e., 24.79 acres. This made it possible to bring large number of peasants under this Act. Under this provision of 24.79 acres the Government acquired only 1.14 lakhs acres of land till 1970-71 as against the estimated available 11.49 lakh acres (Agricultural Census 1970-71) or 8.49 lakh acres as estimated by 26th round of NSS (NSS 26th round July 71 and September 71). Assuming the census figure to be a realistic estimate, the performance was to the tune of only 13.43. This proves the ineffectiveness of the implementation process of the Act. The implementation was very slow and there was no monitoring system. No additional staff were recruited for time bound completion of the project. A Directorate in the name of Directorate of Land Requisition Acquisition and Reforms was established in 1968 to handle the reforms matter especially the ceiling matters. But the Directorate remained more or less inactive. What is surprising is the fact that out of the above acquired land of 1,14,379 acres till 1971 only 15,022 acres or only 13.13 per cent were distributed.

In 1971, the provisions of the Act were further amended drastically. It was amended in tune with the resolution adopted in the Chief Ministers Conference in 1972. The amendment reduced the ceiling limit to 16.53 acres. The Act came into force immediately. At the same time many Indian States have also amended their existing legislations following the recommendations of 1972. The land ceiling limits of some major States are shown in Table 20.1.

(1) It may be observed that the only in respect of three States namely Kerala, Manipur and Tripura, the ceiling limit is lesser than that of Assam in all respects. One notable factor that plays role in this count is the total geographical area and the gross sown area.

Table 20.1: Ceiling Limits in Different States

Sl.No.	Name of the State	Irrigated	Non-irrigated
1.	Andhra Pradesh	10-27 acres	35 to 54 acres
2.	Assam	16.53 acres	16.53 acres
3.	Bihar	15-25 acres	30-45 acres
4.	Gujarat	10-27 acres	20-54 acres
5.	Haryana	7.25-10.9 hect.	21.8 hect
6.	Himachal Pradesh	10-15 acres	30-70 acres
7.	Jammu & Kashmir	8.93-12.5 acres	14.70-22.73 acres
8.	Karnataka	10-30 acres	54 acres
9.	Kerala	12-15 acres	12-15 acres
10.	Madhya Pradesh	18-27 acres	54 acres
11.	Maharashtra	18-36 acres	54 acres
12.	Manipur	5 hect.	6 hect.
13.	Orissa	10-15 acres	30-45 acres
14.	Punjab	7-11 acres	20.5 acres
15.	Rajasthan	18-27 acres	54-175 hect.
16.	Sikkim	12.5-50 acres	12.5-50 acres
17.	Tamil Nadu	12 acres	60 acres
18.	Tripura	4-12 acres	4-12 acres
19.	Uttar Pradesh	7.30 hect.	10.95-18.25 hect.
20.	West Bengal	12.35 acres	17.30 acres

Source: Ministry of Agriculture and Rural Development, 1985.

Except Assam all the three States had agricultural land areas much smaller than that of Assam. Total agricultural area of Kerala (18.02 lakh hect., was less than half than that of Assam (39.40 lakh hectares), while that of Manipur (1.75) was twenty third times lesser and Tripura (3.08 lakh hectares) almost thirteenth times lesser than that of Assam. Thus, these States had their reasons to prefer a lower ceiling limit. But despite having such a bigger area, Assam choose to fix only 16.5 acres as ceiling limit in 1972 which is the lowest when compared to the State areas.

(2) Similarly, the irrigation criterion was also neglected while fixing the limit. Despite the fact that during 1971-72 only 24 per cent (5.72 lakh hectares) of the net areas sown of the State were irrigated the lawmakers preferred to consider the land of the State as best category of land. The Chief Ministers' Conference recommended that "The best category of land in a State with assured irrigation and capable of yielding at least two crops a year should have ceiling

within the range of 10 to 18 acres taking into account of the fertility of the soil and other conditions." Of all the conditions laid down by the conference, our policy makers had preferred to accept only one that is the numerical limits. Had the policy makers examined the prevailing agricultural conditions thoroughly, they would not have accepted the first category rather would have opted for the second category, which was defined as "the land having assured irrigation for only one crop in a year the ceiling shall not exceed 27 acres." Thus, the imposition of ceiling limit at 16.6 acres itself was a big blunder committed by the Government.

Government sat on the newly enacted (1972) amending legislation till 1975. No specific guidelines were issued clarifying as to how to implement the provisions of the Act and amalgamate the pending cases. The Tea Garden land although was brought under ceiling, yet no ceiling limit was fixed for the tea estates. The Act exempted the entire area under tea cultivation from the operation of the Act along with the lands used for ancillary purposes as well as land for future expansion of tea cultivation.

As per record, in 1975 the first comprehensive Government instruction was issued to the Deputy Commissioners or the Collectors, i.e., the implementing authorities of the ceiling Act—covering the details of the methods to be followed while implementing the provisions of the Act.[7] The Circular also dealt to the procedure to be followed while acquiring the tea garden lands. Despite that the acquisition of land did not gain any momentum. The declaration emergency in 1976 had a direct impact on the implementation of the Act. The Government of India took exceptional measures to implement the ceiling laws by bringing it within a time frame. Annual targets were fixed for each State and they were told to achieve the same annually. The observation of late Prime Minister Mrs. Indira Gandhi on pursuing land reforms programme is significant: "I think, it is either because there are people who have vested interest in not having reforms or because of a general lack of appreciation of the vital role of land reforms that there is a certain amount of lethargy and hesitation in implementation.[8]

We would not discuss further details of the Act except that the provisions provided for disposal of the acquired lands. As per the recent available data total returns filed were 21,894 out of which

21,720 were disposed of but still 174 remained to be disposed. On the other hand government have declared 6,13,400 acres of land as ceiling surplus till date. But out of this only 5,75,337 acres were taken possession of. Out of the total areas taken possession of 5,45,871 acres have so far been distributed. The performance appears to be quite satisfactory when it is viewed in the light of performance of other Indian States. It may be observed from Table 20.2 that the position of Assam is seventh with reference to area acquired and area distributed.[9] Jammu and Kashmir occupies the top position distributing 98.78 per cent of the total acquired land followed by Haryana and Orissa. Assam still has 67,530 acres of land to be distributed. Out of this 38,451 acres are involved in judicial courts since 1992. The State Government has done nothing to dispose of these cases.

Table 20.2: Distribution of Ceiling Acquired Lands in Different States of India (2003)

Sl.No.	State/U.T.	Area distributed (in per cent)
1.	Jammu & Kashmir	98.78
2.	Haryana	95.00
3.	Orissa	93.34
4.	Manipur	91.90
5.	Uttar Pradesh	91.27
6.	Tamil Nadu	90.23
7.	Assam	88.99
8.	Maharashtra	86.86
9.	Tripura	80.10
10.	West Bengal	76.89
11.	Rajasthan	75.57
12.	Bihar	73.88
13.	D&N Haveli	72.88
14.	Andhra Pradesh	72.82
15.	Madhya Pradesh	62.57
16.	Gujarat	61.50
17.	Kerala	48.61
18.	Punjab	46.72
19.	Karnataka	46.23
20.	Pondicherry	46.00
21.	Delhi	34.81
22.	Himachal Pradesh	1.95

Source; Department of Land Resources, Government of India.

The performance of the State in respect of distribution of ceiling surplus land since 1998 is negligible. The main defects in the implementation of the Act are the faulty provisions of the Act and rules framed thereunder. At every stage, there are loopholes right from determination of ceiling land through fixation of compensation to selection of landless cultivators. The subordinate revenue officials who are required to prepare the statement of tenure holders having more land than the ceiling limit were not properly trained in their work. Moreover they were sympathetic to the tenure holders and the tenure holders on their part resorted to all means to save their lands from acquisition. There was no definite information about the total land held by individual tenure holders in the State. It was left to the personal knowledge, conjecture and hearsay of the lowest revenue functionaries. As a result most of the tenure holders' land were left out and many were accounted for only partial acquisition. At every stage they were helped by corrupt officials taking advantage of the loopholes in the Act. Of all provisions of the Act, the provision of review was the most obnoxious provision. This provision was laid down in Section 7(6) of the Act where a review against that order of the Collector lies with the Government. Corrupt officials misused this provision several times to suit their own needs. The Assam Land Reforms Commission while examining the provisions of the Act and in respect of certain cases in a district commented, "Government order passed in revision allowing additional area of 140 acres, therefore violate the legal provisions of Rule 5(A). The Commission is shocked to find out such deliberate violation of the provision of the Act."[10]

The vesting of the surplus land in the State is the first step towards subsequent action in implementing the objectives of the Act. But in most cases possession of the acquired land took place only in black and white. No physical possession was taken over. This happened especially in respect of tea gardens. The tea gardens are continuing to occupy the ceiling surplus land and utilising for their benefits. As per Government estimate as much as 9,256 acres of ceiling surplus land is under the occupation of tea gardens.

The Effect

Our experience in the implementation of the Act has been unique in the country. The Ceiling Act had inevitably led to further fragmentation of the holding, which were already small and uneconomic. Undoubtedly such further fragmentation was against the interest of the State. One of the objectives of the Ceiling Act was to ensure greater production by reducing the area of big farms and distributing small patches of land to landless peasants. The theoretical arguments in favour of small farms is that with high yielding varieties, availability of cheap fertiliser and irrigation facilities, production is bound to increase and even very small holdings can be economic. This theory has also been advocated by David and Herz on the above assumptions. The most unfortunate is the fact that it was earlier stated that the majority of the land holdings in the State were small holdings. As per the NSS 17th round, the State in 1961-62 had as much as 38.73 per cent marginal operational holdings (less than 1.01 hectares) and 37.94 per cent small holdings (1.01 to 2.02 hectares), which together constitutes 76.67 per cent of the total holdings (NSS report of 1961-62, 17th Round). The strength of these marginal and small holdings increased to 61.57 and 24.32 per cent in 1980-81 (NSS Report 1980-81, 37th Round). This shows the gravity of the situation. There was no effort to consolidate the areas neither any endeavour was made to form cooperative farming of small farmers. In the Land Policy of 1972, Government made desperate attempt to form Agricultural Farming Corporation which finally was a total failure due to lack of motivation of the farmers.

A situation of this type was not assessed earlier. The marginal farms did not anyway rehabilitate the small peasants. But at that time the sole aim of all including the peasants were to acquire the ownership rights of the small holdings that they were cultivating. At the political level, it was considered desirable to distribute the acquired surplus land to maximum number of people irrespetive of the size of the land allotted to him and the economic benefits accruing from the operations. The up-to date statistics shows that altogether 2,36,893 acres of land was distributed among 1,79,208 beneficiaries implying thereby that at an average of only 1.32 acres of land per family was distributed. Thus another few lakhs of uneconomic holdings in the State were created. Some of these holdings were converted from agricultural holdings to residential holdings and used for purposes other than agriculture.

Under Section 16 of the Act, it is mandatory on the part of the Government to offer settlement to the cultivating tenants within a prescribed period on realisation of 15 times the land revenue. But in actuality, there emerged great difficulty in identifying the tenants, as the tenancy operation conducted during 1972 have left out from recording many cultivating tenants and in some cases draft tenancy records were issued but due to hurried closure of the operation, final records were not issued. As a result, most of the tenants were deprived of their rights of getting title over the lands they cultivate.

One of the objectives of the ceiling Act was to ensure greater production by reducing the area of big farms and distributing the acquired land to the landless peasants. The idea behind this was that these small and marginal farmers will put all their family labour in the upkeep and maintenance of the farm as well as for better supervision of crop and protection from pests, drought and insects. But the available statistics do not establish this. Rice production in Assam in 1961 was 2,474 thousand tonnes, which in ten years reached 3,266 in 1972 a rise of 32.00 per cent. It was the time when the full benefit of the Ceiling Act was yet to be reaped. Unfortunately during the peak years of the implementation of Ceiling Act when the ceiling benefits were available to the people, i.e., during 1975 and 1985 production rose from 3,475 thousand tonnes to 4,271[11] thousand tonnes giving an increment of only 22.90 per cent. This means that diminishing returns had already started operating in the field. Thus even this dream of higher productivity was not met.

Thus to conclude, the major factor that impeded the basic land reforms measures in the State are:

1. Faulty provisions of the law.
2. Untrained land records staff on which the whole State was to depend for identification of the land.
3. Corrupt officials who encouraged handing over of the barren and inferior land as ceiling surplus retaining the best.
4. Failure of the Government to pursue the cases in judicial courts.

NOTES

1. Government of Assam, Department of Agriculture, 2002.

2. Report of the Land Reforms Committee, 29, 1956, Government of India.
3. Nehru, Jawaharlal, *Autobiography.*
4. Report of the Land Reforms Committee, 1956, Government of India.
5. Implementation of Land Reforms—A Review by the Land Reforms Implementation Committee of NDC, PC, 1966, 34.
6. NSS Report (8th Round) 1953-54, No. 66.
7. Handbook of Government Circulars, Vol. I.
8. Proceedings of Chief Ministers' Conference 1976.
9. Government of India, Department of Land Resources, 2003.
10. Report of the Assam Land Reforms Commission, 1981.
11. *1961-69: India Directorate of Economics and Statistics*, All India Estimate of Rice (various issues) 1970-99; Agricultural Centre for Monitoring Indian Economy, *Agriculture* (various issues).

21

MASS MIGRATION OF AGRICULTURAL POPULATION AND ITS EFFECT

A Study in the Laharighat Block

— NILOFAR JASMIN

Introduction

The trends in migration have been considered as a sensitive index of changing pattern of economic opportunities in an area. Low production in an area may be one of the causes of migration and migration may also affect production. Generally, production largely depends upon its skilled and efficient manpower. And if mass migration takes place in an area, particularly of agricultural labourer, it certainly affects the productivity of that area. But while discussing the productivity, lack of irrigation facilities, high yielding varieties of seeds, high techniques, pest management, manure, etc. are mainly taken into account but often overlooked another important factor, i.e., migration.

It was observed that in the Laharighat Block of Morigaon District, Assam, migration is becoming a regular phenomena and the production is also in decreasing trend. Therefore, the present study was carried out purposively in the Block, with a specific objective to find out the causes of mass migration and its effect on agricultural activities and productivity of the area.

Study Area

Laharighat Block—the present study area is one of the four Blocks of Morigaon District, Assam. The district is situated on the south bank of river Brahmaputra and is susceptible to flood due to its low lying topography.

The Block comprises of many '*Char*' areas inhabited by the immigrant Muslim people. They are hardwoking and skilled and efficient agriculturists. Once the Block was known as best belt of jute, rice and vegetable production. But during last ten years approximately 23,914 persons have migrated from the area due to different reasons, resulting to the labour crisis problem in the area.

Presently, there are 185 revenue villages in the Block with 2,46,254 population according to 1991 census report.

Materials and Methods

This chapter is based on both primary and secondary data. The secondary data have been collected from S.D.C. Office and Block Office, Laharighat and District Agriculture Office, Morigaon. As block-wise productivity data were not available, the Development Officer, Agriculture of Laharighat and Bhuragaon Circle had provided unofficial approximate data for the same. Census report of 1991 and 2001 were also consulted whenever necessary.

To collect primary information two sets of respondents were selected purposively and interviewed. One set comprises of 200 agricultural labour and marginal farmers migrated from different villages of Laharighat Block to different districts of Assam and neighbouring States. They were interviewed in February 2003 when they came back to their native villages to celebrate Id festival. The second set of respondent comprise of 90 farmers. They were selected from five non-eroded villages, viz., Tengaguri, Utalabari, Gerua, Phalihamari and Bhuyabari. The farmer have been categorised as marginal, small and big farmer.

In both cases only immigrant Muslim male members were selected as subject as in that society women plays no vital role in agricultural activities.

Survey schedule and interview techniques were used besides observation method and informal discussion in the study. And suitable statistical tools were also used for analysis of data.

Results

Table 21.1 shows the number of population migrated from the area.

In last ten years 23,914 individuals migrated from the area of study. Highest migration took place in the year 2000 (20.49%).

Table 21.1: Migration of People from Laharighat Block
(*No. of revenue village: 185; No. of eroded village: 52*)

Year	No. of persons migrated	Percentage %
1991	519	2.17
1992	415	1.72
1993	1,200	5.02
1994	2,040	8.53
1995	2,550	10.66
1996	2,210	9.24
1997	424	1.76
1998	2,224	19.29
1999	3,302	13.80
2000	4,900	20.49
2001	4,140	17.31
Total	23,914	100%

Source: S.D.C. Office, Laharighat & Block Office, Laharighat.
Except the marginal farmers and labourers other persons and farm families are also included in the Table.

Table 21.2 shows the locations to which the people had migrated. Out of 200 persons interviewed 61.0 per cent migrated to Dimapur of State Nagaland, others migrated to different places of Upper Assam, viz., Dibrugarh (5%), Digboi (7%), Sivasagar (6%) and Tinsukia (4%). Three per cent were found to be migrated to Itanagar of Arunachal Pradesh, Uriam Ghat (5%), Umrangsu of Meghalaya (5%) and Garamapani (3%) and Nagaon (1%).

Tables 21.3 and 21.4 show the changing pattern of occupation and the past and present income of the migrants. Forty four per cent migrants were marginal farmers. Others were also engaged mostly in agricultural activities such as agricultural labour (35.5%), seasonal agricultural worker (7.5%), attached or bonded labourer (5.5%). Only 7.5 per cent migrants were formerly engaged in different types of works such as hotel boy, worker in shops, petty vendors of fish and vegetable, etc. In the village, earning of these people were in between Rs. 500-Rs. 1,500 per month.

Table 21.2: Migration and Distance

Places to where migrated	State	No. of individuals	Percentage
Dimapur	Nagaland	122	61
Digboi	Assam	14	7
Tinsukia	Assam	8	4
Dibrugarh	Assam	10	5
Sivsagar	Assam	12	6
Umrangso	Meghalaya	10	5
Garampani	Assam	6	3
Nagaon	Assam	2	1
Itanagar	Arunachal Pradesh	6	3
Uriam Ghat		10	5
	Total	200	100

Source: Personal survey.

Table 21.3: Past Occupation and Income

Past occupation	Monthly income in Rs.	No. of respondents	Percentage
1. Agricultural labourer (Daily wager/monthly L)	800-1,050	71	35.5
2. Bonded or attached agricultural labour (yearly)	50-200	11	5.5
3. Seasonal Agricultural Worker	1,100-1,500	15	7.5
4. Marginal Farmer	—	88	44
5. Others (Workers in shop, hotels, vendor, etc.)	600-900	15	7.5
Total Respondents		200	100

Table 21.4: Present Occupation and Monthly Income

Occupation	No. of respondents	Percentage	Monthly wage (in Rs.)
1. Construction	153	76.5	2,000-3,200
2. Daily labourer	21	10.5	2,000-2,300
3. Agricultural workers	10	5.5	1,500-2,500
4. Truck loading—unloading	7	3.5	1,800-2,500
5. Rickshaw puller	4	2	1,500-1,800
6. Vendors (Vegetable, fish, oil, egg, etc.)	5	2.5	1,200-1,800
	200	100	

Source: Personal survey.

The present income of the migrants is in between Rs. 1,200 to Rs. 3,200 per month as shown in Table 21.4. Income is higher in construction works and 76.5 per cent migrants work in construction sites and roads. Others have engaged themselves in different types of works such as daily labourer in factories, private houses, and coal fields (10.5%), truck loading and unloading (3.5%), rickshaw puller (2%), vendors, hotel boy, etc. (2.5%) and 5.5 per cent migrants works in the agricultural field in Nagaland with Naga landlord.

Table 21.5 shows the number of family members of both sets of respondents. Most of the families consist of 7-9 family members (55.5%), 31.03 per cent consist of 4-6 members. 10.63 per cent of 10-12 members and 1.72 per cent of 13-15 members while only 1.03 per cent family consists of 2-3 members.

Table 21.5: Size of the Family

Type of family	No. of family members	No. of respondents	Percentage	
Small family	1-3	3	1.03	
Large family	4-6	90	31.03	86.53
	7-9	161	55.5	
Very large family	10-12	31	10.03	11.75
	13-15	5	1.72	
Total No. of respondents		290	100	

Source: Personal survey.

In Table 21.6, the opinion of the migrants are highlighted about the problems in their native villages, which led them to migrate. Sixty per cent migrants expressed that job opportunity is not adequate in the village, 35 per cent migrants were mainly marginal farmers having lost their land and property in erosion. And 5 per cent individuals migrated because they are attracted to urban life.

Table 21.6: Causes of Migration shown by the Respondents

Causes	No. of respondents	Percentage
Inadequate job opportunity	120	60%
Erosion and flood	70	35%
Urban life attached	10	5%
	200	100

Source: Personal survey.

Table 21.7, assesses the constraints faced by the existing farmers in the village of the Laharighat Block because of migration of agricultural labourer. Except the marginal farmers others had to leave the land fallow. (2 to 11 *bighas*) due to labour crisis in each season particularly in summer paddy, jute and vegetable cultivation.

Table 21.7: Constraints Faced by the Farmers

Category of farmers	No. of farmers interviewed	Land left without farming	The crops
Marginal farmers (Less than 7.5 *bighas* of land)	30	Nil	
Small farmers (7.5-15 *bighas*)	30	2 to 5 *bighas*	Bodo, Jute, vegetable, wheat
Big farmers (more than 15 *bighas* of land)	30	2 to 11 *bighas*	Bodo, jute, vegetable, ahu, wheat
Total	90		

Source: Personal survey.

Table 21.8, shows the productivity trend of different crops for last ten years. It is seen that productivity of most of the crops are in decreasing trend.

Discussion

Analysis of data presented in Tables 21.1 to 21.7 reveals that the rate of migration is increasing almost steeply from 1991 to 2001. It appears that Dimapur is the first choice of the migrants of Laharighat Block. And then they prefer to go to different industrial towns and the construction sites where job is assured and wage is high.

Interestingly, though farmers of the Block are facing labour crisis, most of the migrants are complaining of non-availability of job in the village. It is because job related to agriculture is not full time job and population pressure in the area is very high (as found from the analysis of Table 21.5). Therefore job is not available in proportionate to human resources. Moreover the wage in the village is very low. So most of the poor people like to migrate for better income.

Table 21.8: Productivity Trend of Different Crops

Year	Place	Crops AY						Vegetable AY		
		Winter rice	Summer rice	Autumn rice	Jute	Mustard	Wheat	Cole Crops	Tomato	Brinjal
1991-92	L	1,950	3,200	744	3,580	750	1,300	2,180	2,570	2,750
	M	1,622	2,420	1,200	3,220	520	1,283	—	—	—
1992-93	L	1,750	2,800	590	3,440	720	1,228	2,090	2,580	2,700
	M	1,429	2,317	1,104	3,028	489	1,178	—	—	—
1993-94	L	1,880	2,800	590	2,267	756	1,224	2,160	2,588	2,888
	M	1,624	2,063	103	2,192	389	757	—	—	—
1994-95	L	—	2,610	540	1,730	650	1,050	2,000	2,460	2,650
	M	1,196	1,417	856	2,018	393	731	—	—	—
1995-96	L	1,356	2,450	544	1,956	625	1,050	1,980	2,400	2,350
	M	1,160	2,102	656	1,831	393	915	—	—	—
1996-97	L	1,210	2,400	.....	1,650	565	1,089	1,850	2,356	2,550
	M	1,297	1,823	644	1,929	570	9,29	—	—	—
1997-98	L	1,290	2,580	.....	1,630	450	1,200	1,660	2,210	2,275
	M	1,207	2,226	793	1,785	396	1,536	—	—	—
1998-99	L	1,189	2,250	....	1,345	450	1,000	1,580	1,880	2,120
	M	1,154	2,590	590	1,796	352	1,324	—	—	—
1999-2000	L	1,360	2,450	.....	1,343	342	680	1,440	1,650	2,145
	M	1,144	2,200	590	1,706	398	590	—	—	—
2000-2001	L	1,250	2,350	.....	1,250	310	675	1,320	1,589	2,098
	M	1,056	2,008	560	1,650	325	500	—	—	—

AY = kg/hectare AY = Average yield L = Laharighat M = Morigaon — = Data not available ..... = Negligible production

Source: District Agriculture Office, Morigaon.

Further, regular occurrence of the flood even twice or thrice in a single year, massive erosion and heavy siltation on cultivable land are becoming a regular phenomenon in the Block as a consequence of construction of Kaliabhomora Bridge in 1984-85. Already 53 villages have been completely eroded in the last 10 years affecting 48,000 farm families of the area.

Considering the low trend of productivity of *Kharif* crops due to heavy monsoon and consequent flood, farmers of the area mainly depend upon the *Rabi* crops (crops of flood based season). This cropping pattern requires a large number of labourers. But due to labour crisis, presently, most of the farmers are bound to leave a portion of land without cultivation and thereby affecting the productivity of the area.

Conclusion

From the above discussion we may come to the conclusion that the whole problem is a very complicated one. Regular occurring of flood, massive erosion, migration and problems of low production are inter woven with each other. Population pressure, low wage, inadequate job opportunity and urban oriented mentality are the main causes of migration.

This chapter clearly indicates that the migration has a direct effect on production because in flood free season farmers cannot cultivate the entire land because of crisis of labour which aggravates the problem of production.

REFERENCES

Chandna, R.C. (2001). *Geography of Population*, Kalyani Publishers, Ludhiana.

Hussain, M. (2002). *Human Geography*, Rawat Publication, Jaipur & New Delhi.

Kaur, G. (2002). "Rural to Rural Male Migration in India, 1991", *The Transactions*, Vol. 24, Nos. ! & 2, 71-83.

Ray S. (2003). "Rural-Urban Migration: A Case Study of a Squatter Settlement", *Journal of Indian Anthropological Society*, 38 (55-59), 2003, pp. 55-59.

22

WOMEN IN AGRICULTURE
A Case Study of the Mishing Tribe in Dhemaji District of Assam

— M. GOSWAMI, P.J. SHARMA, B.S. MIPUN and P.C. SARMA

Introduction

Women bear the primary responsibility for actions needed for education, nutrition, health and family income. Over 70 per cent of India's population currently derive their livelihood from land resources, which includes a sizeable proportion of economically active women. Women's contribution to agriculture, when measured in terms of the number of tasks performed and time spent, is always substantial. Female work participation rate (FWPR) is defined as the proportion of female workers to the total female population. FWPR is highly correlated with the level of technology, occupational pattern and social groups particularly among the weaker sections of the society.

Agriculture is the dominant sector of the economy of Assam contributing about 56 per cent of the State's income (Das, 1984) and crop production occupies the most important part of agriculture. Agriculture in turn is dependent upon type of soil, rainfall and host of other natural factors. Moreover, the cultivable land is highly limited due to the peculiar physiographic conditions of the State. Intensity of cropping and yield per hectare are also very low in the State.

The rural women of Assam have a long tradition of doing various economic and productive activities although their role has been visualised as a passive one. Since independence, various socio-economic changes have taken place in the rural society of Assam due to various developmental activities under successive planning. The impact of these changes has been noticed on their attitude and outlook. The changing role of rural women on the productive activities are also visible.

Assam is the homeland of many ethnic groups with different languages, customs and with different historical, economical, cultural traditions and religious beliefs. Although, most of them share a common mode of income from agricultural activities, even than there exists variation in the pattern of agricultural practices among different groups. Female work participation in this sector is also of varied nature and among tribal population the work participation of female is always higher than the non-tribal population.

Assam lies in the North-Eastern region of India and has a unique geographical condition as it is surrounded by hills in every direction and receives a very high annual rainfall combined with very high humidity creating a favourable atmosphere for various primary activities. Dhemaji district is located in the eastern most district of Assam and is the homeland of a large number of tribes, namely the Mishings, the Deoris, the Sonowal-Kacharis, the Bodos, the Tiwas besides other ethnic groups like the Chutiyas, the Ahoms and others. Agriculture is the main occupation of all the tribes.

The Mishings constitutes the bulk of the tribal population of the district and are mainly riverine dwellers. The tribe constitutes 21 per cent of the total population of the district. In the Mishing culture, men are highly regarded and play a dominant role as the heads of the household. On the other hand, women enjoy very low status and are considered inferior to men. As such, most of the women are lowly educated and are usually made to marry at a tender age. The tribe depends mainly on agricultural farming and fishing as their means of livelihood. In addition, they rear domestic animals such as pigs, cows, goats, buffaloes and fowls. A large proportion of female workers are engaged in the agricultural activities and perform certain activities which are generally not done by the males, such as transplantation, manual threshing, dehusking, etc.

In the light of the above discussion this chapter aims to explore the differential rate of participation of tribal women in agricultural activities with the following objectives:

1. To study the personal and socio-economic characteristics of the Mishing women.
2. To make a comparative study of the association between personal and socio-economic characters of the tribal women and contribution to family economy through their rate of participation in crop production.

Methodology

The present study is based on the data collected from the secondary sources, which ultimately have been authenticated by the primary data collected personally in the field in the form of interview and scheduled questionnaires. Primary data collected from five Mishing villages (selected from the dominant population in the villages with more than 80%) of the district through scheduled questionnaires. To make a comparative study five non-tribal populated villages were also sampled in a similar manner. Ten respondents were selected at random from each village to make a final sample size of 100. The source of secondary data is district census handbook, departmental records and other published articles. A geographical study of the environmental resources pertaining to agricultural crops and animal husbandry necessitated special reference to topography, soil type, climate and irrigation facilities.

Study Area

Dhemaji district is located in the eastern most part of Assam. The district has two sub-divisions and 5 revenue circles with 1,205 inhabited villages. The area of the district is 3,237 sq km with density of population of 148 persons per sq km. A substantial population in the district is constituted by the tribals (43.92%). The major tribes of the district include the Mishings, the Sonowal Kacharis, the Bodos and the Deoris. The district ranked 11th in literacy rate (65.96%) among the districts of Assam. The district has a fair proportion of women engaged in cultivation and agricultural activities (32.9%).

The district lies between Arunachal Pradesh in the north and in the east, river Brahmaputra in the south and the Lakhimpur district in the west. The district is situated between 94° and 95°3′ East longitude and 27°25′ and 27° 55′ North latitude.

The economy of Dhemaji is generally agro based. Sericulture, fishing and driftwood business is practiced in smaller scale. However, sand deposition and other adverse effects of chronic floods on fertile agriculture land have made even affluent farmers landless. Lack of good communication system, shortage of power and lack of proper irrigation and marketing facilities add to the poverty of the district.

The types of soil of Dhemaji district vary according to the topography. The soil is generally sandy to alluvial and clay loam. Forest occupies 19.67 per cent and agriculture is done on 39.48 per cent of the area. The mean temperature ranges between 10°C and 35°C, while the mean annual rainfall is around 2,626 mm. The rainfall is distributed throughout the year except for the months of November to January. In the summer months high rainfall is accompanied by flash floods and high floods, which besides destroying houses, destroys standing crops and disrupts the communication systems. Such fertile land turns into barren sand plots and became unfit for cultivation. In fact flood affected 28,084 hectares, 1,31,475 hectares and 56,432 hectares of cropped area in the years 1998, 2000 and 2001 respectively. The annual flood and silting by the river Brahmaputra is mainly responsible for making the district one of the poorest in Assam.

Agriculture and Agro-biodiversity

A traditional method of cultivation with a very low degree of agro-biodiversity is practiced in Dhemaji district of Assam. The net sown area is 55,821 ha. The major portion of this land has rainfed cultivation. About 6,233 ha of land have irrigation facilities and of these only 87.5 ha are utilised. Much of the land have medium to low water holding capacity suitable for rainfed or irrigated cultivation of paddy, pulses and vegetables.

Most of the traditional crops are photosensitive and thus the sowing time is the most important variable in determining the yield. The practice of inter-cropping, multiple cropping, mixed cropping and crop rotation are almost absent in the district. During the agricultural season the entire household is involved in agriculture in this area. Mishing women are involved right from selection of seeds to transplanting, weeding, threshing and winnowing. During the preparation of the land, the men folk go early to the fields while the women do their household chores including cooking, cleaning, fetching water and caring of the children. Womenfolk then go to the field carrying food for the men. In the field, they assist in breaking the clods, building bunds, etc. Women's agricultural work takes place largely from July to November. Crops grown here are primarily rainfed. Paddy is the main crop. Cash crops and vegetables are grown in only a few isolated pockets of the district. The first peak period of work in agriculture is June to August, when women engage in breaking up sods of earth, transplanting and weeding. The second peak period is October through November, when women are involved in harvesting, drying, pounding/threshing and dehusking paddy. Participation of women in planting and weeding is significantly higher than that of men. The post-harvest work of women includes drying, cleaning and storing the grains.

The tribes residing in the district have simple indigenous methods of selecting seeds and protecting them from insects and pests. The harvested grains are carefully cleaned by winnowing. During the winnowing process, the off-colour and undersized grains are removed. The selected grains are then sun-dried for two or three days and stored as seeds. Seeds are also selected by collecting healthy ear heads with well-formed grains during the harvest and kept aside for the subsequent season's planting. Women play a major role in this process.

Analysis of the Case Study Area

The study of spatial distribution in work participation rate in the State indicates that this rate is high in the tribal dominated and economically backward districts like Lakhimpur (42.86%), Karbi Anglong (42.78%), etc. It might be said that the economic conditions and social prejudices play an important role in determining the female work participation in different parts of the State. Women contribute considerably to household income through farm and non-farm activities as well as through work as landless agricultural labourers. The Indian work force participation rate is 37.7 per cent (Census, 2001) the rate of women is 22.7 per cent which is lower than half the rate of man (51.6%). The work participation rate of female in the district is 34.80 per cent, which is much higher than State average (18.09%). The percentage of female main workers to total female population in Dhemaji district is 17.18 per cent against 10.52 per cent for the State of Assam. The corresponding figures in female marginal workers are 17.62 per cent and 7.57 per cent in Dhemaji and Assam respectively. This indicates that most of the female workers in the district are underutilised. The proportion of main (Agricultural and Other) workers and marginal workers in the district is depicted in the Table 22.1.

Table 22.1: Distribution of Main and Marginal Workers in Dhemaji District of Assam

Workers category	Total	% to total population	% to gender population	% to total workers
Main Workers				
Agricultural Workers				
Total	1,36,164	28.44	—	63.24
	(44,04,081)	(19.65)		(54.45)
Male	91,417	19.09	36.79	42.45
	(35,53,156)	(15.85)	(30.48)	(43.93)
Female	44,747	9.35	19.42	20.78
	(8,50,925)	(3.80)	(7.91)	(10.52)
Other Workers				
Total	24,669	5.15	—	11.45
	(25,87,975)	(11.55)		(31.99)
Male	22,113	4.62	8.90	10.27
	(20,87,031)	(9.31)	(17.90)	(25.80)
Female	2,556	0.53	1.11	1.19
	(5,00,944)	(2.23)	(4.66)	(6.19)
Marginal Workers				
Total	54,494	11.38	—	25.31
	(10,96,879)	(4.89)		(13.56)
Male	5,977	1.25	2.41	2.78
	(1,24,213)	(0.55)	(1.07)	(1.54)
Female	48,517	10.13	21.06	22.53
	(97,266)	(0.43)	(0.90)	(1.20)

Note: Figures in parentheses indicates the All Assam averages.
Source: Statistical Handbook of Assam, 2002.

The proportion of population engaged in economic activities (Work Participation rate) and the occupational structure of the works indicate to a great extent the overall economic development scenario of a region. Within the State of Assam the work participation is considerably higher in the rural areas (36.71%) than that of urban areas (30.92%). The main economic activities of the tribal population in Assam lie in the primary sector of agriculture. In rural areas including the Dhemaji district of Assam, agriculture and allied industrial sector employ most of the total female workers. Fig. 22.1 shows the comparative statistics of workforce of non-tribal and tribal women of the district. For the purpose villages were selected in such a way so that the ethnic population is 60-80 per cent and above 80 per cent.

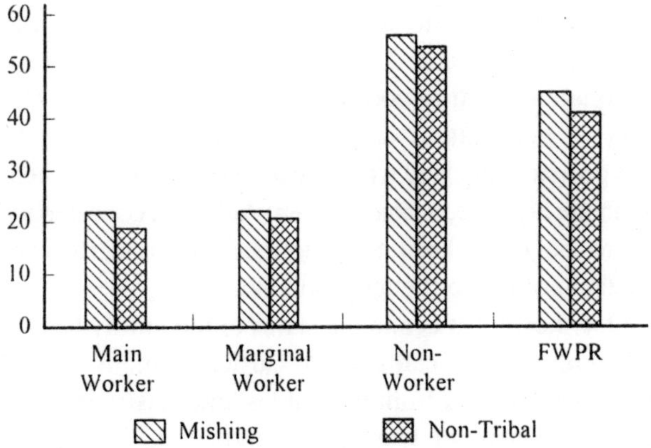

Fig. 22.1: Per cent of Female Workers in Villages having Ethnic Population more than 80 per cent.

This figure indicates that the work participation rate of tribal women is higher than the non-tribal population in the same geo-climatic and agro-economic condition prevailing in the district.

Demographic features of female population like age composition, age at marriage, health care, fertility and child care practices, rural-urban composition of population, nature of migration determine the potential labour force which can be utilised for productive purposes. Social status, customs, attitudes and religious beliefs affect the volume of female labour supply in the society. Educational standard also influence the nature of occupation of females. Table 22.2 and illustrations show the relationship of socio-economic characteristics of Mishing women with their work participation rate in terms of number of days spent in agricultural activities. For comparison incorporation of the non-tribal component had also been done in Table 22.2 and illustrations.

It is evident from Table 22.2 that the percentage of workers of Mishing women is substantially higher than non-tribal women irrespective of age group. The percentage of non-workers (3.33%) of Mishing women reflects the higher FWPR of the tribe as compared to 31.67 per cent of non-workers in case of non-tribal women. Table 22.3 shows that the women remained economically active upto 45 years of age after which their economic contribution tends to decline. However, the economic contribution of women workers in case of non-tribal women is found to be higher than their tribal counterpart.

It is observed from Table 22.3 that educational status of women and their economic contribution towards family economy through participation in agricultural activities are inversely linked. The illiterate and primary educated women participate more in agricultural activities than the higher educated women. Although educated women are known to be more keen to assimilate the modern trends in agricultural technology for better production.

Table 22.4 shows that economic status of the family has a direct relationship with the women's work participation in agriculture. Further it is also observed from the Tables and illustrations that the non-worker component of women respondents is higher in non-tribal population. Thus it is clear that the female work participation rate of Mishing women has always been higher than the non-tribal women of the area irrespective of their age, education and land holding.

Conclusion

The agriculture scenario in Dhemaji district is characterised by a very low degree of agro-biodiversity with the practice of traditional methods of cultivation. It is clear that the Dhemaji district is no longer a closed tribal system. It is an agricultural system moving from subsistence to semi-commercial agriculture. It is a society marked by the breakdown of local self-sufficiency and moving towards increasingly more inclusive and complex systems.

Table 22.2: FWPR in Agriculture in relation to Age

No. of working days	Percentage of respondents and their economic contribution to total family income (in percentage) according to age group							
	Below 30 years		Between 30-45		Above 45		Total	
	% of respondents	Economic contribution	% of respondents	Economic contribution	% of respondents	Economic contribution	% of respondents	Economic contribution
Mishing Women								
Non-workers	1.67	0	—	—	1.67	0	3.33	0
Below 100	28.33	6.29	28.33	7.38	15.00	6.26	71.67	6.64
Between 100-150	1.67	19.09	11.67	21.72	—	—	13.33	20.40
Above 150	1.67	21.13	10.00	23.81	—	—	11.67	22.47
Total	33.34	46.51	50.00	52.91	16.67	6.26	100.00	49.51
Non-tribal Women								
Non-worker	11.67	0	1.67	0	18.33	0	31.67	0
Below 100	10.00	12.44	18.33	8.37	23.33	8.45	51.67	9.75
Between 100-150	1.67	22.97	6.67	22.17	5.00	16.31	13.33	19.31
Above 150	3.33	23.70	—	—	—	—	3.33	23.70
Total	26.67	59.11	26.67	30.54	46.66	24.76	100.00	52.76

Table 22.3: FWPR in Agriculture in relation to Education

No. of working days	Percentage of Respondents and their economic contribution to total family income (in percentage) according to the level of education									
	Illiterate		LP		High school		HSLC and above		Total	
	% of respon-dents	Economic contri-bution	% of respon-dents	Economic contri-bution	% of respon-dents	Economic contri-bution	% of respon-dents	Economic contri-bution	% of respon-dents	Economic contri-bution
Mishing Women										
Non-worker	—	—	—	—	1.67	0	1.67	0	3.33	0
Below 100	48.33	6.80	5.00	5.66	8.33	6.94	10.00	7.15	71.67	6.64
Between 100-150	10.00	20.16	1.67	21.04	—	—	1.67	20.00	13.33	20.40
Above 150	8.33	22.64	1.67	31.07	1.67	13.74	—	—	11.67	22.47
Total	66.66	16.53	8.34	19.26	11.67	6.89	13.33	9.05	100.00	12.38
Non-tribal Women										
Non-worker	20.00	0	1.67	0	8.33	0	1.67	0	31.67	0
Below 100	28.33	10.93	5.00	9.81	15.00	10.13	3.33	8.13	51.67	9.75
Between 100-150	8.33	18.36	3.33	20.45	1.67	19.09	—	—	13.33	19.31
Above 150	—	—	—	—	3.33	23.70	—	—	3.33	23.70
Total	56.66	9.76	10.00	10.09	28.33	13.23	5.00	4.07	100.00	13.19

Table 22.4: FWPR in Agriculture in relation to Land Holding

No. of working days	Percentage of respondents and their economic contribution to total family income (in percentage) according to Land Holding							
	Below 2 acres		Between 2-3 acres		Above 3 acres		Total	
	% of respondents	Economic contribution	% of respondents	Economic contribution	% of respondents	Economic contribution	% of respondents	Economic contribution
Mishing Women								
Non-worker	—	—	—	—	3.33	0	3.33	0
Below 100	1.67	6.31	—	—	70.00	6.97	71.67	6.64
Between 100-150	13.33	20.41	—	—	—	—	13.33	20.41
Above 150	5.00	25.96	6.67	18.98	—	—	11.67	22.47
Total	20.00	52.68	6.67	18.98	73.33	6.97	100.00	49.52
Non-tribal Women								
Non-worker	3.33	0	8.33	0	20.00	0	31.66	0
Below 100	10.00	11.65	8.33	9.80	33.33	7.80	51.66	9.75
Between 100-150	5.00	20.91	5.00	20.09	3.33	16.93	13.33	19.31
Above 150	3.33	23.70	—	—	—	—	3.33	23.70
Total	21.66	56.26	21.66	29.89	56.66	24.73	99.98	52.76

REFERENCES

Das, M.M. and Dutta, L. (1986). "Regional Variation in Land Use and Agriculture in North East India", *North Eastern Geographer*, 18(1-2), pp. 21-25.

Das, R. (1979). *Women's Education in Assam in Post-independence Period (1947-71) and Its Impact on the Social Life in the State*, Unpublished Ph.D. Thesis, Gauhati University.

Desai, S. (1993). "Gender Inequalities and Demographic Behaviour in India", *The Population Council*, Inc., New York.

Jain, D. and Chand, M. (1982). *Report on a Time Allocation Studies: Its Methodological Implications*, Paper presented at Technical Seminar on Women's Work and Employment, New Delhi.

Kar, B.K. (2002). *Women Population of Northeast India: A Study of Gender Geography*, Regency Publications, New Delhi.

Mazumdar, V. (1975). "Women in Agriculture", *Indian Farming*, 25(8), pp. 5-7.

Nayak, D.K. (1991). *Female Participation in Economic Activity in Selected Rural Areas in India: A Geographical Analysis*, Unpublished Ph.D. Thesis, Jawaharlal Nehru University, New Delhi.

Raju, S. (1993). "Regional Disparities in Female Literacy in Urban India: Problems and Prospects", in Nuna S.C. (ed.), *Regional Disparities in Education Development*, South Asian Publishers Pvt. Ltd., New Delhi, pp. 243-65.

Singh. R.D. (1996). "Female Agricultural Workers' Wages, Male-female Wage Differentials, and Agricultural Growth in a Developing Country, India", *Econ Dev Cult Change*, 45(1), pp. 89-123.

Usha, M.S.S., Singh, A.J. (1983). "Female Labour Participation in Faridkot District (Punjab State)", *J Res. Punjab Agriculture University*, 20(4), pp. 525-535.

23

ECONOMICS OF POTENTIAL AGRO-HORTICULTURE SYSTEMS IN MID HILLS OF MEGHALAYA

— S.B. SINGH, S.V. NGACHAN and L.K. MISHRA

Introduction

Shifting cultivation, locally known as *jhuming*, is a predominant system of farming in Meghalaya. About 265 thousand ha of land in the State is affected by shifting cultivation and 52 thousand families are dependent on it for their livelihood (Government of India, 1995). Basically, the system consists of felling and burning of patch forest, cultivating for one to two years and abandoning it for several years to enable the forest to regenerate (Toky and Ramakrishnan, 1981). Shifting cultivation in its more traditional and culturally integrated form, is ecologically and economically viable system of agriculture as long as population densities are low and *jhum* cycles are long (Anon, 1992). However, due to population pressure on land, *jhum* cultivation becomes uneconomical and unsustainable causing drastic decline in crop yield, loss of forest wealth, soil fertility, biodiversity and environmental degradation (Moursi, 1984; Christenty, 1986). In Meghalaya, the average annual loss of top soil, organic carbon, P_2O_5 and K_2O due to shifting cultivation were to the extent of 40,900 kg, 702.9 kg, 0.15 kg and 7.5 kg per ha (ICAR, 1983). Recognising the evil effects of shifting cultivation, Indian Council of Agricultural Research (ICAR) Complex for North-Eastern Hill (NEH) Region have developed different agro-horticulture systems of farming as alternatives to shifting cultivation. Keeping the above facts in view, this chapter examines the production potentials and financial viability of selected agro-horticulture systems relative to alternative systems in the State. The results of the study would be very useful to the administrators and policy makers in popularisation of agro-horticulture systems in the State.

Material and Methods

Data Sources

The present study was based on experimental data collected during 1987 to 1999 at the agroforestry farm of ICAR Research Complex for NEH Region, Umiam. The experimental site is located at an altitude of 980 m above msl. Rainfall distribution in this ecology is highly skewed with maximum recorded during June to July with average of 2,235 mm annually. Fertile *alfisols* are the predominant soil type with *p*H of 5-5.2. The Agro-horticulture systems studied consisted of two fruit trees namely, mandarin and guava in combination with different crop rotations, i.e., soybean, groundnut, turmeric and ginger. In case of turmeric, the crop was grown continuously for 5 years followed by frenchbean in 6th, 7th, 8th and 9th year and again turmeri in 10th, 11th and 12th year. Leguminous crop, frenchbean was included in the crop rotation to inbuilt soil fertility. Similar crop rotation was followed for ginger. The fruit trees were planted on hill slopes after terracing with spacing of 5 m × 5 m. Therefore, altogether 400 plants each of mandarin and guava could be accommodated in one ha area. Area occupied by intercrops in each system was approximately, 8,400 sq.m. No intercrops were grown in the first year due to land designing and plantation of fruit trees. All the intercrops included in the experiment were also grown in nearby farming system fields to assess their productivity in open without fruit tree canopies. In order to test the difference between the mean yields on monocropped and agro-horticulture plots '*t*'-test was applied.

Financial Analysis

The cost streams involve establishment (fixed) cost and recurring/ maintenance cost. The former includes costs on field clearing and land designing, digging and filling of pits, seedlings and soil treatment, etc. The latter covers expenditure on manure, fertilisers, insecticides, pesticides, weedicides, weeding and pruning and costs incurred on seasonal crops. The benefit streams include all the direct benefits from fruits and seasonal crops. Net Present Value (NPV), Benefit-Cost (B-C) Ratio, Payback Period and Internal Rate of Return (IRR) have been computed for financial analysis. Similar methodology was used by Shanmugam, 1988 and Nair and Dagar, 1991 in financial analysis of agroforestry systems in other parts of India. A discount rate of 18 per cent was used in presenting results and the inputs and outputs were valued at the market price of 1999-2000. The NPV, B-C Ratio, IRR and Payback Period were defined as:

$$\text{NPV} = \sum_{t=1}^{n} \frac{(B_t - C_t)}{(1+r)^t}$$

$$\text{B-C ratio} = \frac{\sum_{t=1}^{n} \frac{B_t}{(1+r)^t}}{\sum_{t=1}^{n} \frac{C_t}{(1+r)^t}}$$

$$\text{IRR} = \sum_{t=1}^{n} \frac{(B_t - C_t)}{(1+r)^t} = 0$$

where t = Years 1, 2, 3, ..., n.

B_t = Benefit from the system in period 't'

C_t = Cost associated with the system in period 't'

r = Discount rate, and

n = Number of years used in the analysis.

Internal Rate of Return (IRR) is the rate of discount which equalises the present worth to the total investment and is calculated by equating NPV = 0. Payback Period is the time period for the agro-horticulture systems to pay back its investment and is calculated by determining the value of n which satisfies the equation of IRR = 0.

Results and Discussion

Effects of Fruit Species on Crop Yield

Inspection of monocropped soybean, groundnut, turmeric and ginger yield in Table 23.1 shows they varied very little between 10.15 per cent in soybean to 16.38 per cent in ginger. By contrast, when these crops were grown over the 12 years study period as components of agro-horticulture systems, the variation range from 20.03 per cent in guava-soybean to 34.01 per cent in guava-groundnut. The '*t*'-test results also showed significant variation between the monocropped and agro-horticulture plot yields. This is because their yields, which initially demonstrated the expected supplementary of intercropped systems, declined/inclined with time reflecting competition/complementary from the developing over-storey of fruit trees. Clearly, the introduction of fruit tree species did not reduce row crop production through competition during the initial 3 years. From 4th year onwards, all row crop yields were reduced indicating the effects of competition with the fruit tree component. By 4-6 years soybean production of agri-horti plots was reduced to 78 per cent or less of the monoculture yields. Groundnut yields from agri-horti plots for 4-6 years were also reduced, they ranged from 63 to 68 per cent of

Table 23.1: Influence of Fruit Species on Yield of Intercrops (Q/ha) During 12 Years in Meghalaya State

Systems	Yield (3 years average)			Mean yield	Coefficient of variation	*t*-value
	1-3	4-6	10-12			
M+soybean	18.99	14.63	12.81	14.94	24.32	2.38**
G+soybean	16.21	13.71	12.73	13.92	20.03	2.79**
Monocropped soybean	19.19	18.52	18.45	18.39	10.15	—
M+groundnut	24.94	14.00	10.01	19.37	24.32	2.80**
G+groundnut	19.57	13.20	9.88	13.39	20.03	3.21***
Monocropped groundnut	20.78	20.70	20.15	20.90	10.15	—
M-turmeric	163.08	118.50*	216.33	171.91	28.79	3.16***
G-turmeric	152.38	109.50*	150.00	140.77	24.63	1.85*
Monocropped turmeric	14.57	112*	125	131.51	13.45	—
M-ginger	98	68.96*	110.34	95.59	29.88	2.44**
G-ginger	98.43	70.51*	102.00	92.79	26.57	2.19**
Monocropped ginger	99.87	65.68*	86.81	86.43	16.38	—

***, **, * indicate significant at 1, 5 and 10 per cent levels respectively.
Note: M and G denote mandarin and guava respectively.

the monoculture yields. In 10-12 years, soybean yield on mandarin and guava plots were 69.4 per cent and 69.3 per cent of monoculture yields. For groundnut it was 49.7 per cent and 49 per cent respectively. It clearly indicates significant reduction in yields of soybean and groundnut after 4th year of cultivation. But for ginger and turmeric, their production in agri-horticulture plots decline in 4-5 years aş lower than proportionate than monocrop yields. It is attributed to depletion of soil fertility because of continuous cultivation of rhizomous crops rather than over-storey of fruit trees. To improve fertility status of the ginger and turmeric plots, leguminous crop of frenchbean was grown continuously from 6th to 9th year. Production of turmeric and ginger on intercrop plots increase more than proportionately than monocrop yields at 10-12 years. It clearly shows that for rhizomous crops like turmeric and ginger, there is no competition with the fruit trees but complementary relationship exists between them over the years.

Effects of Competition on Fruit Yields

Table 23.2 shows that average yield from sole fruit trees appeared to be greater than the comparable agro-horticulture systems. However, none of the '*t*' values in Table 23.2 was found to be significant. These data suggest that the fruit yields in the agro-horti plots were not affected by competition. In fact, the increasing yields of turmeric and ginger on agro-horti plots without reducing fruit yields, indicated a complementary relationship.

Table 23.2: Effect of Inter-crops on Yield of Fruit Trees in Meghalaya State

Land-use system	Mean yield (Qt/ha)	Coefficient of variation	*t*-value
Mandarin (sole fruit)	215.67	52.82	—
Mandarin + soybean	210.27	55.32	1.38^{NS}
Mandarin + groundnut	212.01	61.12	1.80^{NS}
Mandarin + turmeric	205.03	58.25	1.50^{NS}
Mandarin + ginger	201.35	70.87	1.78^{NS}
Guava (sole fruit)	49.32	75.62	—
Guava + soybean	44.04	70.87	0.89^{NS}
Guava + groundnut	45.03	68.35	1.70^{NS}
Guava + turmeric	42.01	72.41	1.14^{NS}
Guava + ginger	43.02	70.30	1.38^{NS}

Note: NS indicates non-significant.

Financial Analysis

The costs of growing mandarin-based and guava-based agro-horticulture systems are given in Tables 23.3 and 23.4 respectively. The total establishment cost for both the systems was found to be same, i.e., Rs. 38,960 per ha. Contour terrace preparation formed the major share of about 69 per cent of establishment cost followed by pitting and planting (21%) and seedlings (10%) costs. The average annual cost of practicing mandarin based agro-horticulture systems were worked out to be Rs. 31,210 for M+soybean, Rs. 32,896 for M+groundnut, Rs. 45,746 for M+turmeric and Rs. 44,814 for M+ginger (Table 23.3). Human labour constituted a dominant share, which ranged from 35 per cent in Mandarin-turmeric to 47 per cent for Mandarin-soybean. Seed input was found to be equally important for M+turmeric and M+ginger systems. For guava-based systems, the average annual cost ranged from Rs. 29,948 in G+S to Rs. 43,658 in G+turmeric systems (Table 23.4). Similar resource use patterns were found for guava based systems.

Table 23.3: Average Cost of Production of Mandarin-based Agro-horticulture Systems in Meghalaya State (Rs./ha)

Inputs	Mandarin + soybean	Mandarin + groundnut	Mandarin + turmeric	Mandarin + ginger
Fruit seedlings	308	308	308	308
Seed of crops	1,047	1,454	10,132	8,478
Chemical fertilisers	5,061	5,061	6,059	6,800
Farmyard manure	5,631	5,631	7,492	7,492
Insecticides + pesticides	1,495	1,510	1,602	1,555
Soil treatment with chemicals	38	38	38	38
Human labour	15,180	15,900	15,956	16,066
Interest on working capital @ 10%	2,837	2,991	4,159	4,074
Total cost	3,120	32,896	45,746	44,814

Note: Establishment cost was worked out to be Rs. 38,960 for all the four systems.

Results of financial analysis of different systems showing net present value (NPV), benefit cost ratio (B/C), internal rate of return (IRR) and Payback period (PBP) are presented in Table 23.5.

Table 23.4: Average Cost of Production of Guava-based Agro-horticulture Systems in Meghalaya State (Rs./ha)

Inputs	Guava + soybean	Guava + groundnut	Guava + turmeric	Guava + ginger
Fruit seedlings	308	308	308	308
Seed of crops	1,047	1,453	10,132	8,478
Chemical fertilisers	4,544	4,544	5,542	6,283
Farmyard manure	5,169	5,169	6,654	6,877
Insecticides + pesticides	966	1,005	1,105	1,095
Soil treatment with chemicals	38	38	38	38
Human labour	15,152	15,457	15,914	16,094
Interest on working capital @ 10%	2,723	2,798	3,969	3,917
Total cost	29,948	30,777	43,658	43,086

Note: Establishment cost was worked out to be Rs. 38,960 for all the four systems.

Financial Return without Fruit Trees

Ginger provided the highest return with a NPV of Rs. 1,56,689 followed by turmeric (Rs. 1,55,682). The least was groundnut with NPV of Rs. 5,610. Benefit cost ratio was also highest for ginger (1.81) and minimal for groundnut (1.05). Similar pattern was followed for IRR and payback period.

Financial Return with Fruit Trees

All the systems analysed under this category were profitable at 18 per cent discount rate. Dividing the systems into fruit trees and agro-horticulture components, a pattern is very evident in the financial indicators. The NPVs and B-C ratios for the agro-horticulture systems were always higher than those for fruit components. Mandarin+ginger system worked out to be most profitable with benefit-cost ratio of (2.58) followed by mandarin-turmeric (2.54) and mandarin-soybean system (2.48). Mandarin-groundnut system being least profitable with B-C ratio of (2.35). The NPV varied from Rs. 2,27,854 in mandarin-groundnut to Rs. 3,71,449 in mandarin-ginger system. The Payback period criterion also indicated whole cost of the systems could be recovered within 1 to 2 years in case of mandarin plus

Table 23.5: Financial Indicators of Selected Agroforestry Systems in Meghalaya State

Indicator	Sole fruit	Mandarin inter-cropped with				Sole fruit	Guava inter-cropped with			
		Soybean	Groundnut	Turmeric	Ginger		Soybean	Groundnut	Turmeric	Ginger
(1)	(2)	(3)	(4)	(5)	(6)	(7)	(8)	(9)	(10)	(11)
NPV (Rs. in '000)	195.86	23.89	227.85	369.54	371.44	15.36	66.15	59.89	1,63,316	198.03
B-C ratio	2.28	2.48	2.35	2.54	2.58	1.16	1.43	1.38	1.75	1.87
IRR (%)	36	42	40	83	105	3.2	45	42	99	130
Payback period (years)	8.77	7.33	7.47	1.69	0.63	3.91	4.17	4.21	1.68	1.22

(Contd.)

Table 23.5: (*Contd.*)

Indicator	Agricultural crops			
	Soybean	Groundnut	Turmeric	Ginger
(1)	(12)	(13)	(14)	(15)
NPV (Rs. in '000)	15.36	5.61	155.68	156.68
B-C ratio	1.16	1.05	1.74	1.81
IRR (%)	32	22	94	104
Payback period (years)	3.91	4.70	2.17	1.58

ginger and turmeric systems whereas it takes 7 to 8 years in case of mandarin plus soybean and groundnut systems. Similarly, in case of guava based systems, guava-ginger was found to be most profitable and guava-groundnut system to be least profitable. Comparison of the two systems showed that mandarin based systems was more profitable from economic point of view, indicated by high B-C ratio and NPV. The lower Payback period in case of guava based system is due to earlier fruiting of guava as compared to mandarin. But as a whole, mandarin-ginger system showed most profitable as reflected by the financial indicators. As shown by low NPV, B-C ratio and IRR indicates that cultivation of guava as a sole fruit is not profitable in the State. It is attributed to low productivity due to insect and pest problems. The feasibility analysis for the various agro-horticulture systems are encouraging and advocate the advantages of adopting agro-horticulture practices to enhance farm income.

Conclusion

The preceding results show that selected agro-horticulture systems may provide significantly higher productivity and greater economic return to farmer than traditional monoculture systems. None of the fruit trees had a significant adverse effect on crop yield during the first three years. However, yield of soybean and groundnut were adversely affected on fourth year reflecting competition from the developing over storey of fruit trees. For ginger and turmeric, supplementary relationship existed first then exhibited complementary relationship later on. Due to fertility depleting nature of ginger and turmeric, these crops should not be grown continuously for more than three years. Leguminous crops like soybean and groundnut can be included in the crop rotation for improving the soil fertility. The results of financial analysis showed that all the systems analysed were profitable at 18 per cent discount rate. The net present values (NPVs) were positive for all the systems examined. Cost benefit ratios were more than unity in all the cases. But as reflected by the NPV and B-C ratio, mandarin-ginger agro-horticulture system showed most profitable output. Payback period which was quite long (8.77 years) for sole mandarin fruit also reduced to a little less than one year. However, even though the NPV and B-C ratio of mandarin-ginger system is high, the net benefit is negative for the first year because of high establishment cost. Given the subsistence incomes of farmer in the steep slopes of Meghalaya, farmers cannot afford even temporary income reduction. Thus, farmers are unlikely to adopt the system on their own initiative. Therefore, there is need for the government to provide incentive mechanism to encourage agro-horticulture system.

REFERENCES

Anon (1992). *Agro-Climatic Planning for Agricultural Development in Meghalaya*, Working Group, Zonal Planning Team, Eastern Himalayan Region, AAU, Jorhat, p. 145.

Christanty, L. (1986). *Traditional Agriculture in Southeast Asia*, Westview Press, Boulder, p. 323.

Government of India (1995). *Basic Statistics of North-Eastern Region*, Ministry of Home Affairs, North Eastern Council Secretariat, Shillong, p. 210.

Indian Council of Agricultural Research (1983). *Shifting Cultivation in North East India*, Publication Unit, ICAR, New Delhi, p. 68.

Moursi, A.W.A. (1984), *The Future of Shifting Cultivation in Africa and the Task of Universities*, Food and Agriculture Organisation (FAO), Rome, p. 198.

Nair, P.K.R. and Dagar, J.C. (1991). "An Approach to Developing Methodologies for Evaluating Agroforestry Systems in India", *Agroforestry Systems*, 16: 55-87.

Shanmugam, K.R. (1998). "Methodologies for Optimal Agroforestry Systems", *Productivity*, 39(2): 337-42.

Toky, O.P. and Ramakrishnan, P.S. (1981). "Cropping and Yields in Agricultural Systems of the North-Eastern Hill Region of India", *Agro-Ecosystems*, 2: 127-32.

24

AGRICULTURAL SCENARIO IN THE NORTH-EAST WITH SPECIAL REFERENCE TO MIZORAM

— SUKHENDU MAZUMDER

The North-East India is abundant in natural resources with a tremendous potential of growth in its agricultural base which can successfully combat the growing need of the fast growing population in the region. In the hills, it is the age-old traditions and customs that determine the use of land system. However care should be taken that the interest of the future generation is not endangered by eroding ecological assets while bringing in interdependence between ecotechnology and sustainable agricultural development. It is to be understood here that the sustainability is not limited in the sphere of produce but it encompasses socio-economic and ecological parameter also.

It is the natural phenomenon here in the North-East that the entire area is more or less hilly. Hence, holdings are getting smaller and smaller and it is likely to get decreased further as the pressure of population is on the increase. But the crux of the problem is that the practice of shifting cultivation is very much a part of life of the people living in the hilly terrains. In fact, the shifting cultivation breeds a lot of problems which the people must take care of. To be specific it leads to extinction of flora and fauna, a threat to food production in the future, forest denudation, salinity and desertification, etc. It is estimated that the shifting cultivations has affected 4.35 million hectares of forest cover which accounts for almost 5.78 per cent of the forest area. Apart from shifting cultivation, there are factors which cause soil erosion, heavy rainfall, cultivation of crops along the slope and land slide, etc. Soil erosion in the hills having 50-60 per cent slopes in the first year and second year *jhum* and bamboo forest has been estimated at the tune of 147, 170 and 30 tonnes/ha respectively.

Irrigation

The striking point of our agriculture is that it is mostly rain-based and the irrigation facilities are comparatively limited and are far below the agriculturally advanced States. The major sources of irrigation in the North-Eastern Region are perennial streams and tanks. The streams are usually blocked at suitable points and thus water is diverted from terrace to terrace. This system of irrigation is seen in the Kohima district of Nagaland, in the entire State of Sikkim and Meghalaya and in Apatani plateau of Arunachal Pradesh. Recently we notice the introduction of canal irrigation system in the Nagaon district of Assam and in the valley of Manipur. In Jaintia hills of Meghalaya bamboo drip irrigation system is a common practice to be seen.

There is no doubt that North-Eastern Region as a whole receives maximum rainfall in the country and here is the paradox that it suffers more than the other regions in the dry season because of the fact that 90 per cent of the area is hilly where irrigation is quite a difficult task. For example, Meghalaya receives highest rainfall in the country and yet it suffers from acute drought condition in the winter. Hence, the topography and agro-climatic conditions are to be taken care of. In Meghalaya, there is deficit of foodgrains for 1.22 lakh tonnes and the State requires to intensify its multiple cropping and make better use of other inputs such as high yielding variety of seeds, fertilisers and plant protection chemicals, etc.

Organic Farming and the Modern Methods

The clear advantage of organic farming is manifold since it helps controlling the pollution factors including the degradation of environment and above all—the sources viz., vermicomposts, biofertilisers, etc. are quite cheaper which the farmers can use easily. When there are sufficient quantity of earthworms in a field, they enrich the soil by supplying sufficient manures, provided, they are taken care of by providing food for them. Organic farmers do not simply leave their fields to nature, they have to use knowledge, techniques and products available to work with nature. In fact, the earth should be made free of chemical poisoning to supply poison-free health nourishing food for all. Through the organic farming, recycling crop residue in soil will multiply soil organisms and would create soil humus and a pest-resistant soil climate. And in the economic front, the organic farming will help to curtail valuable foreign exchange and thus the import of petroleum products for producing harmful chemical fertilisers and pesticides will be minimised. Since the time of independence, our agriculture has become chemical-oriented and the use of chemical fertilisers has created an acceleration in the growth of plants and the resultant weakening of the tissues and thus necessitating more use of pesticides for the protection of the crops. And in this way the soil has been rendered lifeless too. The vicious cycle does not end here rather it leads to hardening of the soil structure and creation of inorganic chemical which destroys the micro-organism and earthworms and again to combat the same there has been application of increased dose of chemical inputs.

In the North-East there is no dearth of green manures and the same can be used with a clear advantage unlike other advanced States. Thus there will be sustainable agricultural growth as there will be socio-economic and ecological balance too. However the organic manures required per hectare of land is double (e.g., vermicompost) than that of chemical fertilisers. But the benefit derived from the use of organic manure is many times more than that of chemical fertilisers.

In Tripura, rice is the main staple food of the people and here too like that of Meghalaya, the State is deficient in rice requirement. Since 1985 the deficit is at the tune of 1.80 lakh MT a year. The State is facing population explosion and that is impediment to all growth process and thus requirement is increasing gradually. The population rose from 20.53 lakhs in 1981 to 32 lakhs in 2001. The foodgrain deficiency is a big problem in the State and the gap between the requirement and production is in fact is the outcome of poor profitability by the farmers. Hence, there should be proper planning so that the cultivators find the job profitable. There should be better support price which may provide incentive and production may increase. In fact, rice flow from other States are sold at cheaper rates than the local rice and thus rice production is less attractive.

Mizoram Scenario

The State has been divided into six agriculture districts, viz., Champhai, Lunglei, Saiha, Kolasib, Mamit, Aizawl (Fig. 24.1).

The people practise shifting (*jhuming*) cultivation throughout the length and breadth of the State. The main hurdle to the development of agriculture is the want of water during the dry season which in turn does not permit double cropping. However a lot of changes are witnessed these days with the introduction of tractors and power tillers. The conventional system of cultivation is gradually taking the back seat. The farmers are using improved variety of seeds and the rainfed fields are gradually looking for irrigated paddy fields. But the condition of irrigation was not that good prior to the Sixth Five Year Plan as there were very few irrigated fields and they were mostly limited to some temporary arrangement done by the collective efforts of the people in the shape of some small earthen channels and some temporary diversions, etc. Till March 1993 there were 5 lift irrigation schemes and 26 irrigation schemes and they were in the plain areas where rice is cultivated (wet rice cultivation—WRC). But the same has been changed a lot by the year 2000-2001 when the total number of flow irrigation schemes was 7,702 and that of lift irrigation 216 (Table 24.1).

In Mizoram too rice is the main staple food of the people and there has been constant efforts towards the direction to improve the same but the State has still miles to go since almost 50 per cent deficit remains to be met. The total requirement of the State is 14,61,335 qtl and the total production of the State is only 7,35,000 qtls, that is a deficit of 50 per cent (or say 50.30%). Hence to attain self sufficiency in production the State needs to produce additional 7,26,335 qtl (Table 24.2). This is indeed a big task on the part of the State to achieve at the earliest. However there has been constant efforts and resultant improvement in the same when we take the data available from the DAOs and Statistics section in respect of paddy and other food crops including vegetables (see Tables 24.3 to 24.6). However, the average land holdings of the marginal farmers, small farmers, medium farmers and large farmers have gradually declined

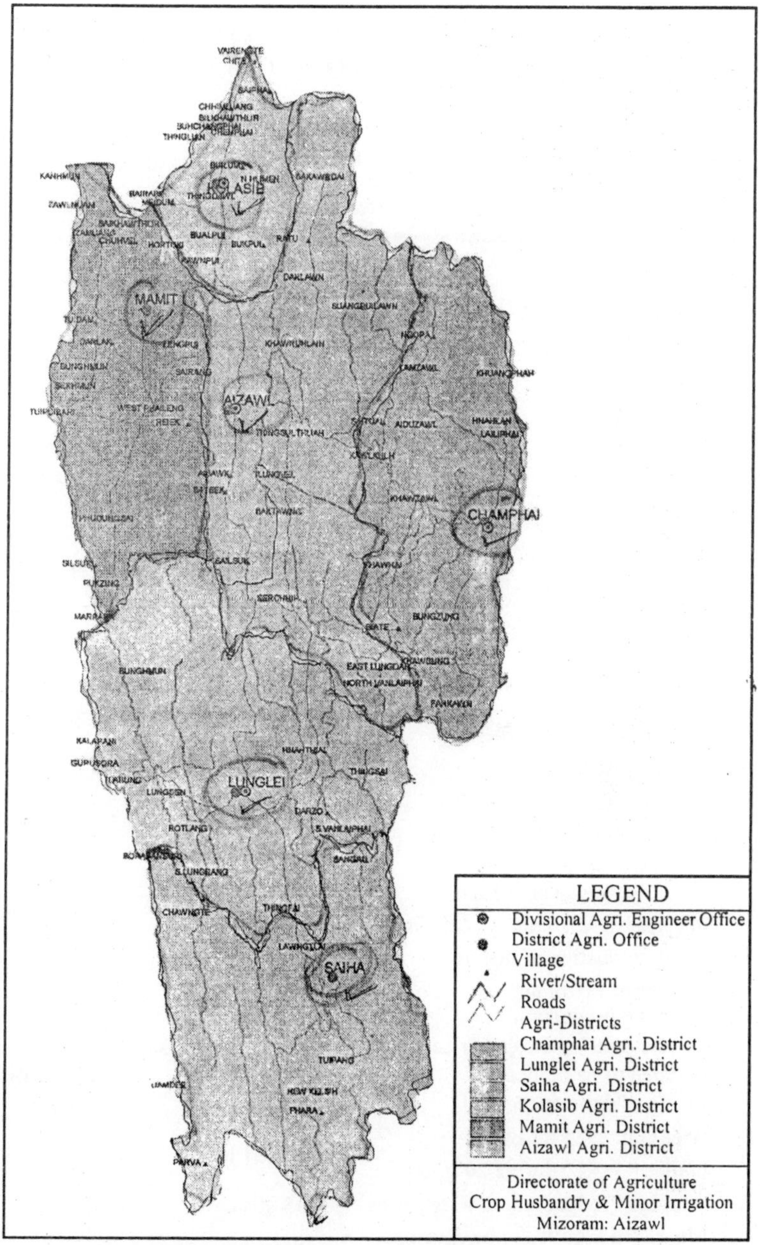

Fig. 24.1: Mizoram showing Agriculture District

from 37.389 (in '000 ha) in 1994-1995 to 25.080 in 2001-2002, 26.705 to 20.065, 32.046 to 25.794 and 46.280 to 34.392 ha respectively (Table 24.6).

Table 24.1: Mizoram: Report on Third Minor Irrigation Census 2000-2001

Sl.No.	District	No. of block	No. of villages	Total No. of schemes					Total	Village schedules
				Dugwell	Shallow	Deep	S. flow	S. lift		
(1)	(2)	(3)	(4)	(5)	(6)	(7)	(8)	(9)	(10)	(11)
1.	Aizawl	5	111	0	0	0	1,213	0	1,213	111
2.	Champhai	3	86	0	0	0	2,667	17	2,684	86
3.	Kolasib	1	25	0	0	0	431	27	458	25
4.	Lawngtlai	2	134	0	0	0	955	0	955	134
5.	Lunglei	4	171	0	0	0	681	0	681	171
6.	Mamit	3	55	0	0	0	287	2	289	55
7.	Saiha	2	75	0	0	0	419	0	419	75
8.	Serchhip	2	30	0	0	0	1,049	170	1,219	30
	Total	22	687	0	0	0	7,702	216	7,918	687

Source: Department of Agriculture, Statistics Section.

Table 24.2: How Far Mizoram is Self-Sufficient in Rice Production

1.	Population of Mizoram (2001 census)	— 8,91,058
2.	Average requirement of Rice per person per year at the rate of 450 g per day	— $\frac{450 \times 365}{1,000} = 164$ kg
3.	Total Requirement of Rice for one year in Mizoram	— $\frac{8,91,058 \times 164}{100} = 14,61,335$ qtls
4.	Total production of rice by farmers of Mizoram	— 7,35,000 qtls
	i.e., % of production	— 50.30%
5.	To attain self sufficiency rice required to be produced	— 14,61,335 – 7,35,000 = 7,26,335 qtls
	i.e., % of deficiency	— 49.70%
6.	Area cultivated for paddy (*Jhum* + WRC) during 2001-2002	— 55,881 ha
7.	Average Productivity of Rice per ha	— 13 qtls/ha
8.	More land required to be developed for self sufficiency in Mizoram	— 55,872 ha

Source: Department of Agriculture, Statistics Section.

Table 24.3: Final Area and Production of Important Crops of Mizoram State for the Years 1998-1999 to 2000-2001

Sl. No.	Name of Crops	1998-1999		1999-2000		2000-2001		2001-2002	
		A	P	A	P	A	P	A	P
1.	Paddy								
	(a) *Jhum*	46,634	58,849	36,285	53,930	35,798	59,560	40,306	63,568
	(b) W.R.C.	16,814	38,332	10,361	29,281	12,953	34,885	12,690	33,845
	(c) H.Y.V.	4,944	13,008	3,067	5,186	3,088	9,228	2,885	8,302
	Sub-Total	68,392	1,10,189	49,713	88,397	51,839	1,03,673	55,881	1,05,715
2.	Maize	8,681	16,589	5,277	9,702	6,643	16,436	7,378	16,646
3.	Wheat	55	44	43	38	—	—	10	18
4.	Pulses	5,346	9,306	4,043	5,741	2,374	3,960	2,677	3,799
5.	Tapioca	486	2,545	351	2,515	—	—	303	1,480
6.	Oil Seed	7,413	7,731	7,294	6,314	7,375	5,682	7,872	5,499
7.	Cotton	555	427	584	350	482	210	62	215
8.	Tobacco	437	434	434	392	373	246	293	226
9.	Sugarcane	1,258	7,863	783	1,550	952	6,361	1,380	9,360
10.	Horti Vegetables	8,163	26,827	—	—	—	—	—	—
11.	Potato	—	—	530	3,924	310	2,429	376	1,472
	Sub-Total	32,394	71,766	19,339	30,526	18,509	35,324	20,356	38,715
	Grand Total	1,00,786	1,81,955	69,052	1,18,923	70,348	1,38,997	76,237	1,44,430

A = Area (in ha); P = Production (in MT)

Source: Department of Agriculture, Statistics Section.

Table 24.4: Survey of Cultivated Area and Production of Paddy in Mizoram during 2001-2002

Agriculture District	Aizawl 'E'	Aizawl 'W'	Lunglei	Kolasib	Saiha	Total
WRC 2001-2002						
Number of Household	43,430	39,565	22,695	16,365	21,185	1,43,240
Area of WRC cultivated (in ha)	6,494	531	621	4,219	2,917	14,782
Number of WRC families	5,293	625	821	2,171	828	9,738
Expected yield from WRC (in MT)	16,567	1,261	1,353	14,062	1,287	34,620
Jhum 2001-2002						
Number of *Jhum* families	21,603	10,062	9,709	3,930	2,707	48,031
Jhum area (in ha)	10,509	15,860	9,774	4,139	9,865	50,147
Expected yield from *jhum* (in MT)	18,917	13,352	11,275	6,784	8,191	58,519
Area of crops cultivated other than Paddy	3,031	5,487	1,563	991	9,521	20,593
No. of families cultivated other crops except Paddy	15,990	8,119	2,785	1,445	4,340	32,679
Fallow land from 1996-2000	41,834	51,378	38,402	15,539	36,776	1,83,929

Source: District Agriculture Officers.

Table 24.5: Estimated Area and Production of Agriculture Crops in Mizoram 2003-2004 (District-wise)

Sl.No.	Name of crops		Aizawl		Mamit		Kolasib		Champhai	
			Area (in ha)	Production (in MT)	Area (in ha)	Production (in MT)	Area (in ha)	Production (in MT)	Area (in ha)	Production (in MT)
(1)	(2)		(3)	(4)	(5)	(6)	(7)	(8)	(9)	(10)
1.	Paddy									
	(1) *Jhum*	(a) Early	4,580	8,700	1,977	3,932	5,500	10,450	100	195
		(b) Late	7,354	15,440	1,184	2,340	1,365	2,661	4,514	9,479
	Total of (a + b)		11,934	24,140	3,161	6,272	6,865	13,111	4,614	9,674
	(2) WRC (*Kharif*)	(a) Early	182	509	335	630	1,560	3,910	0	0
		(b) Late	1,750	4,900	470	1,084	1,890	4,690	4,750	11,875
	Total of (a + b)		1,932	5,409	805	1,714	3,450	8,600	4,750	11,875
	(3) WRC (*Rabi*)	(a) Early	0	0	11	11	255	705	0	0
		(b) Late	0	0	0	0	0	0	0	0
	Total of (a + b)		0	0	11	11	255	705	0	0
	(4) HYV	(a) Early	0	0	91	246	2,380	7,045	0	0
		(b) Late	85	275	7	21	253	758	12	36
	Total of (a + b)		85	275	98	267	2,633	7,803	12	36
	Total of *Jhum*		11,934	24,140	3,161	6,272	6,865	13,111	4,614	9,674
	Total of WRC & HYV		2,017	5,684	914	1,992	6,338	17,108	4,762	11,911
	Grand total of Paddy		13,951	29,824	4,075	8,264	13,203	30,219	9,376	21,585

(*Contd.*)

Table 24.5: (*Contd.*)

Sl.No.	Name of crops		Lunglei		Saiha		Total (Mizoram)	
			Area (in ha)	Production (in MT)	Area (in ha)	Production (in MT)	Area (in ha)	Production (in MT)
(1)	(2)		(11)	(12)	(13)	(14)	(15)	(16)
1.	Paddy							
	(1) *Jhum*	(a) Early	1,728	3,798	2,522	2,469	16,407	29,544
		(b) Late	5,442	12,080	6,010	6,303	25,869	48,303
	Total of (a + b)		7,170	15,878	8,532	8,772	42,276	77,847
	(2) WRC (*Kharif*)	(a) Early	53	159	72	112	2,202	5,320
		(b) Late	448	1,334	1,280	2,248	10,588	26,131
	Total of (a + b)		501	1,493	1,352	2,360	12,790	31,451
	(3) WRC (*Rabi*)	(a) Early	61	183	0	0	327	899
		(b) Late	59	177	0	0	59	177
	Total of (a + b)		120	360	0	0	386	1,076
	(4) HYV	(a) Early	8	24	0	0	2,479	7,315
		(b) Late	8	24	0	0	365	1,114
	Total of (a + b)		16	48	0	0	2,844	8,429
	Total of *Jhum*		7,170	15,878	8,532	8,772	42,276	77,847
	Total of WRC & HYV		637	1,901	1,352	2,360	16,020	40,956
	Grand Total of Paddy		7,807	17,779	9,884	11,132	58,296	1,18,803

Table 24.6: Average Land Holding Size of Farmers in Mizoram

Year	Average Land Holding Size								
	Marginal (1 ha)		Small (1-2 ha)		Medium (4-10 ha)		Large (above 10 ha)		
	Number ('000)	Area ('000 ha)	Number ('000)	Area ('000 ha)	Number ('000)	Area ('000 ha)	Number ('000)	Area ('000 ha)	Total farmers
1994-1995	62.316	37.389	17.804	26.706	5.341	32.046	3.560	46.280	89,021
1995-1996	55.688	33.412	15.911	23.866	4.773	28.638	3.182	41.366	79,554
1996-1997	66.044	39.626	18.869	28.303	5.660	33.960	3.773	49.049	94,346
1997-1998	57.782	34.669	16.506	26.414	4.953	34.671	3.302	42.926	82,547
1998-1999	54.264	27.134	15.505	23.256	4.652	27.912	3.101	37.212	77,522
1999-2000	51.095	25.548	14.599	20.438	4.376	26.256	2.887	34.664	72,190
2000-2001	52.154	26.077	14.901	20.861	4.470	26.820	2.980	35.760	74,506
2001-2002	50.160	25.080	14.332	20.065	4.299	25.794	2.866	34.392	71.658

70% — Marginal farmer — Average area — 0.5 ha
20% — Small farmer — Average area — 1.4 ha
6% — Medium farmer — Average area — 6 ha
4% — Large farmer — Average area — 12 ha

Source: Department of Agriculture, Statistics Section.

Regarding rainfall in the State, the annual average from the year 1986 to 2001 is 2,557 mm (Tables 24.7 and Fig. 24.2). Obviously the total quantum of rainfall in the State is not poor and from the months of May to Sept. It varies from 259 mm to 587 mm (Fig. 24.2). Still Mizoram is known for its water scarcity. The people have been practising roof-top harvesting for long but the same is far below the requirement.

Table 24.7: Year-wise Rainfall Report of Mizoram (in mm) 1986-2000

Sl.No.	Year	No. of Station	Total rainfall collected	Average annual rainfall
1.	1986	13	33,941	2,610
2.	1987	14	35,064	2,504
3.	1988	14	37,479	2,677
4.	1989	14	35,805	2,557
5.	1990	21	55,038	2,621
6.	1991	21	49,374	2,351
7.	1992	21	48,705	2,319
8.	1993	22	59,223	2,691
9.	1994	13	25,605	1,969
10.	1995	13	66,670	2,590
11.	1996	14	35,806	2,557
12.	1997	14	38,956	2,782
13.	1998	14	37,371	2,669
14.	1999	24	62,421	2,600
15.	2000	26	74,952	2,883
16.	2001	26	65,908	2,535

Note: Average Rainfall of Mizoram : 2,557 mm.
Source: Department of Agriculture, Statistics Section.

Apart from rice, there is good prospect of maize production since the same is the second most important cereal crop and is next to rice occupying 8,681 ha and a production 16,589 MT in 1998. However the same is below the average productivity at the national level. The demand is on increase as the same is abundantly used for pig feed, poultry feed and cattle etc.

To conclude, we can say that by tapping the rain water in the slopes, using improved seeds, applying organic fertilisers and also applying chemical fertilisers in a judicious manner, e.g., according to the soil conditions, the agricultural scenario can be improved to a great extent in Mizoram as well as in the whole of North-East.

Table 24.8: Land-Use Statistics of Mizoram—1998-1999, 1999-2000, 2001-2002 and 2002-2003

(*Thousand hectares*)

Heading	1999-2000	2000-2001	2001-2002	2002-2003
I. Geographical Area	2,108.700	2,108.700	2,108.700	2,108.700
II. Reporting Area for Land Utilisation Statistics (1 to 5)	2,108.700	2,108.700	2,108.700	2,108.700
1. Forest	1,598.700	1,626.475	1,626.475	1,541.608
2. Not available for cultivation (a + b)	16.540	130.390	130.490	132.390
(a) Land put to non-agricultural uses	5.200	121,790	122.690	124.190
(b) Barren and unculturable land	11.940	8.600	7.800	8.200
3. Other uncultivated land excluding fallow land (a + b + c)	175.000	58.410	34.863	33.763
(a) Permanent pastures and other grazing land	22.500	22.500	10.600	9.100
(b) Land under miscellaneous tree-crops and groves not included in net area sown	30.600	30.600	19.163	15.163
(c) Culturable waste	121.190	5.310	5.100	9.500
4. Fallow lands (a + b)	220.461	192.094	199.060	247.069
(a) Fallow lands other than current fallows	170.450	155.809	163.262	205.713
(b) Current fallows	50.011	36.285	35.798	41.356
5. Total cropped area	100.739	104.689	117.812	158.397
6. Area sown more than once	2.740	3.358	3.891	4.527
7. Net sown area	97.999	101.331	113.921	153.870
III. Net Irrigated Area	9.217	9.686	10.219	10.669
IV. Gross Irrigated Area	11.160	11.629	12.162	12.612

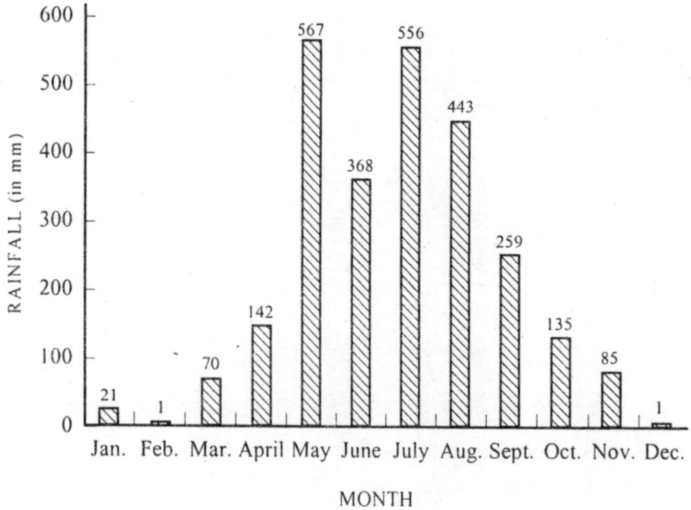

Fig. 24.2: Monthly Average Rainfall, 2002.

REFERENCES

Basic Statistics of NE Region 2000, MHA, Government of India, Shillong.

Government of Mizoram, Agril. Sub-div. Serchhip, abstract of 10 villages, 2002 and 21 villages 2003.

Government of Mizoram, Department of Agriculture (Crop. Husbandry & Minor Irrigation), 17 Oct. 2003.

Journal of NEICSSR, Oct. 2003.

Journal of North-Eastern Council, Jan.-March 2001, Oct.-Dec. 2001, Jan.-March 2002, April-June & July-Sept. 2002, Oct.-Dec. 2002.

Spice India, Vol. 17, No. 1, January 2004.

Statistical Abstract, Department of Agricultural & Minor Irrigation, Mizoram, 2001-2002.

The Telegraph, 6 August 2003, 26 August 2003, 13 June 2003, 7 November 2003.

25

CHANGES IN AGRICULTURAL SCENARIO IN ARUNACHAL PRADESH

— R.P. BHATTACHARJEE

Introduction

Arunachal Pradesh is the largest of North-Eastern States of India with a geographical area of 83,743 sq km. The State has a population of 10,96,702 (Census of India, 2001) which is the sixth highest populated States amongst the eight North-Eastern States. The density of the State is 13 persons per sq km. which is lowest in the country. Thus it is natural to think that agricultural land is more in the State since it is sparsely populated. The State is the eastern stretch of the Himalayas. The forest cover of the State is 82.21 per cent. The forested Siwalik hills rise abruptly to 800 m above the Bramhaputra valley. Most of the ranges of the lesser Himalayas are clothed with temperate forests. The elevation of the hills ranges from 150 to over 7,300 metres. The hill ridges in Arunachal Pradesh are situated in a haphazard manner. As soon as one ridges ends, the other starts. At the intervals, the wide and narrow valleys come into existence. Because of these ridges, the surface of Arunachal Pradesh is found varied almost everywhere which also results in numerous geographical isolations (Bhattacharjee, 2003). Therefore, there is very limited agricultural land. The total operated area of the State is 3,75,892 hectares (Agriculture, 1995-96) which is about 4.5 per cent of geographical area of the State while net area sown is only 2.7 per cent of the reported area (Bhattacharjee, 2000). Thus agricultural changes, if any, will be confined to very limited area of the State because of its large forest area and uneven rugged terrain.

The State has a population of 8,73,633 in rural areas which is 79.66 per cent of its total population. These people in rural areas mainly depend on agriculture. The number of workers in the State is 4,82,206 which consists of 4,13,521 as main workers and 68,685 marginal workers. If we study the categories of workers, we find that 2,81,822 persons are cultivators which is 58.44 per cent of total workers while 18,569 are agricultural labourers. This is 3.85 per cent of total workers. It is observed that these agricultural workers are mainly migrated labourers from plains of Assam consisting of Muslims, Nepalese, etc. Thus we find 62.29 per cent of workers depend on their livelihood directly on agriculture. Thus the main objective of this chapter is to study the changes in agricultural scenario in Arunachal Pradesh.

Changes in Agricultural Scenario in Arunachal Pradesh

The changes in agriculture can be seen in terms of land-use pattern, cropping pattern, cropping intensity, yield-rate, use of modern-technique, settled cultivation and percentage contribution in State Domestic product for Economic Development of Arunachal Pradesh. A few of these are discussed herewith.

Land-Use Pattern

Agriculture in Arunachal Pradesh is influenced by traditional knowledge, customs and festivals. The primitive type of practice of *jhuming* or shifting cultivation is till prevalent in the State. Till the introduction of planned development during early fifties the entire cultivation used to be carried out through *jhuming*. About 4.0 lakh hectares of land was under the coverage of this system. Gradually, as a result of sustained extension effort the extent of such ecological harmful practice has gone down and the present level is about 1.10 lakh hectares (Planning Department, 2002-03). The land-use pattern of Arunachal Pradesh is shown in Table 25.1.

It can be observed that number of holding is increasing in marginal, small and semi-medium size while it is cyclical in medium size and convex at large size. This can be attributed to increase in population, number of families and limited operated area. In the operated area, settled area is increasing while *jhum* type of cultivation

Table 25.1: Land-Use Pattern in Arunachal Pradesh

Particulars	Unit	1970-71	1980-81	1990-91	1995-96	2001-02
1. Operational holding	Nos.	78,990	78,542	93,961	1,06,000	NA
(i) Marginal (below 1.00 ha)		6,060	12,986	16,399	20,400	
(ii) Small (1.00-2.00 ha)		9,450	16,298	17,360	20,500	
(iii) Semi-medium (2.00-4.00 ha)		20,470	22,094	30,202	30,700	
(iv) Medium (4.00-10.00 ha)		28,750	22,073	25,800	28,300	
(v) Large (10.00 ha and above)		14,260	5,091	4,200	6,100	
2. Operated area	Hectare	4,89,011	3,35,732	3,49,891	NA	NA
(i) Settled		28,006	52,012	1,08,803		
(ii) *Jhum*		4,61,005	2,83,720	2,41,088		
3. Average Size of holdings	Hectare	6.19	4.27	3.62	3.55	NA
4. Land-Use Classification						
(i) Net area sown	Hectare	1,15,226	1,18,232	1,65,616	NA	1,99,000
(a) Settled		28,006	52,012	92,616		1,08,500
(b) *Jhum*		87,220	66,220	73,000		90,500
(ii) Current Fallow		51,612	23,127	28,463		
(iii) Other Uncultivated Land (Excluding fallow land)	19,164	34,888	44,437			
(iv) Fallow Land other than current fallow		1,17,740	57,253	35,823		
(v) Cultural Waste land		1,48,698	47,595	32,999		
(vi) Land not available for cultivation		36,571	54,637	42,553		
5. Gross Cropped Area	Hectare	1,29,775	1,36,110	1,93,498	—	2,48,000
6. Area under Multiple Cropping	Hectare	14,549	17,878	27,882	--	49,000
7. Cropping Intensity	Hectare	1.13	1.15	1.17	—	1.25
8. Area under Irrigation	Hectare	22,661	21,857	34,755	—	45,000

Sources: (1) Directorate of Agriculture, *Arunachal Agriculture in Brief, 1980-93*, Naharlogun, pp. 4-5.
(2) Directorate of Agriculture, *Arunachal Agriculture at a Glance, 1993-98*, Naharlogun, pp. 4-5.
(3) Directorate of Economic & Statistics, *Statistical Abstract of Arunachal Pradesh, 2002*, Itanagar, p. 22.
(4) Directorate of Economic & Statistics, *Economic Review of Arunachal Pradesh, 2002*, Itanagar, p. 17.
(5) Directorate of Agriculture, *Agricultural Census, 1995-96*, Series 6, Vol. I, p. 4.

is decreasing but still prevailing in the State. The average size of holdings is also decreasing. During 2001-02, the settled area is shown more than *jhum* cultivation but in the absence of cadastral survey or agricultural census, it is difficult to believe that there is more settled cultivation in the State compared to *jhum* cultivation.

Cultural waste land area is decreasing which indicates that wasteland development programme will be limited in near future. The decrease of current fallows indicates the trend of small *jhum* cycle in the State in *jhum* cultivation. Gross cropped area is increasing and multiple cropping area is also increasing and cropping intensity is also higher as compared to last decades. These are good indicators. The area under irrigation is increasing at a slow pace. But these channel irrigation systems are based on run-off water and these are useful only during rainy season as per observation. And during rainy season, this system of irrigation is of not much use. There is need of large reservoir for any attempt to increase agricultural production through irrigation.

Let us now study the cropping pattern of agriculture in Arunachal Pradesh.

Cropping Pattern

By cropping pattern, we mean the proportion of area under different crops at a point of time. So a change in cropping pattern implies a change in the proportion of area under crops. There are some who believe that cropping pattern is more or less given and cannot be easily changed. They accept everything with a spirit of resignation and sense of fatalism. For them agriculture is a way of life rather than a commercial proposition. Such a highly pessimistic view is not acceptable. Change in cropping pattern is sometimes necessary and this has got to be brought about in the minds of our cultivators for better economic growth (Bhattacharjee, 1998). The cropping pattern in Arunachal Pradesh is shown in Table 25.2.

We find from Table 25.2 that share of foodgrains is decreasing while non-foodgrains is increasing during the period under study. In recent years, the share of fruits is increasing and this is a good change in agricultural scenario in Arunachal Pradesh. Horticulture can usher an era for economic development of the State. Let us now find the changes of yield rates of different products.

Table 25.2: Cropping Pattern in Arunachal Pradesh

Crops	1970-71	1980-81	1990-91	2000-01	2001-02	2002-03
1. Rice	59.2	57.1	49.1	41.0	40.2	40.3
2. Maize	25.7	14.7	15.1	13.3	13.0	13.2
3. Millet	14.9	11.5	8.0	7.0	6.7	6.8
4. Wheat	0.2	1.9	1.4	1.5	1.3	1.3
5. Pulses	NA	0.7	2.2	2.1	2.3	2.4
Total Foodgrains	100.0	85.9	75.8	64.9	63.5	64.0
6. Oilseeds	NA	4.3	9.1	9.1	9.3	9.2
7. Spices	NA	0.9	1.1	3.2	3.3	3.8
8. Sugarcane	NA	0.4	0.1	0.3	0.3	0.2
9. Vegetables	NA	6.2	6.5	7.2	8.1	6.9
10. Fruits	NA	2.4	7.4	15.3	15.5	15.9
Total Non-food	NA	14.1	24.2	35.1	36.5	36.0

Sources: (1) Department of Agriculture, *Arunachal Agriculture in Brief, 1980-93*, Naharlogun, p. 6.
(2) Bhattacharjee, 1998, *ibid.*, p. 38.
(3) Department of Economics & Statistics, *Statistical Abstract, 2001*, Itanagar, pp. 20-30.
(4) Department of Horticulture, *Annual Operating Plan, 2003-04*, Naharlogun.
(5) Department of Agriculture, *Annual Operating Plan, 2001-02, 2002-03 & 2003-04*, Naharlogun.

Yield Rate of Main Agricultural Crops

The yield rate of agricultural crops is determined by the amount of produce per hectare. These are shown in Table 25.3.

From Table 25.3, we find yield rates of rice is more or less stagnant and there is no remarkable improvement in other food products except marginal changes in wheat and pulses. There is noticeable improvement in spices specially ginger in Lohit district and other eastern districts of the State.

We find that there is not much significant change in land utilisation pattern. Land is used mainly for *jhum* cultivation except during the year 2002-03. The modern process of agriculture is not relevant and irrigated area or use of area under manure/fertiliser or HYV/Improved seed is very less. Even the plant protection is also low.

Table 25.3: Yield Rate of Main Agricultural Crops

(Quintals/hectare)

Crops	1970-71	1980-81	1990-91	2000-01	2001-02	2002-03
Rice	13.87	10.57	11.70	10.98	11.26	12.24
Maize	10.19	11.25	12.17	13.56	13.60	13.92
Millet	5.86	8.80	9.78	8.64	9.00	9.20
Wheat	6.45	13.68	19.05	13.01	14.03	15.19
Pulses	—	6.25	7.84	10.12	10.53	10.67
Oilseeds	—	9.40	9.66	9.81	10.63	10.47
Potato	—	71.21	65.41	65.39	70.24	69.63
Spices	—	17.34	23.88	32.31	36.32	30.00

Sources: (1) Department of Agricultural, A.P., *Arunachal Agriculture in Brief, 1980-93*, Naharlogun, pp. 6-13.
(2) Department of Agriculture, A.P., *Annual Operating Plan, 2003-04.*
(3) Department of Horticulture, A.P., *Annual Operating Plan, 2003-04.*

From the statistical abstract of Arunachal Pradesh, 2002 it can be seen that the total consumption of chemical fertiliser was 715 MT and out of that 321 MT was consumed in Tawang and West Kameng district during 2001-02. Similarly during 2000-01, only 712 MT of chemical fertiliser was used in Arunachal Pradesh. HYV seeds has been introduced in the State and yield rate is more but it is not so much popular. The yield rate of HYV rice was 17 quintals per hectare during 2001-02 as compared to 11.30 in total rice yield. The Government has taken a step forward in this direction by the introduction of New Agricultural Policy of Arunachal Pradesh during October 2001. The basic objectives of the Policy are:

1. To increase farmer's income;
2. To address problems related to shifting cultivation;
3. Area specific development strategy;
4. Convergence of allied activities;
5. Technology transfer;
6. Supply of inputs;
7. To facilitate private investment in agriculture;
8. Encourage people's participation;
9. Research and technology package;
10. Enhanced marketing infrastructure;
11. To set up agro-processing units in production;

12. Price support through procurement;
13. Use of Information Technology to ensure greater reliability of estimates in planning and policy making; and
14. Flood and drought management.

The announcement of agricultural policy is a major change for future agricultural development in the State but this will depend on the proper implementation of the policy declaration.

Level of Foodgrains

Population is increasing in the State. It was 4.68 lakhs during 1971 which has increased to 10.97 lakhs during 2001. The foodgrain production has also increased in Arunachal Pradesh but the State is yet to attain self-sufficiency in foodgrain production. The level of foodgrain production is shown in Table 25.4.

Table 25.4: Level of Foodgrain Production in Arunachal Pradesh

(MT)

Crops	1970-71	1980-81	1990-91	2000-01	2001-02	2002-03
Rice	52,662	88,191	1,42,500	1,32,690	1,34,621	1,52,500
Maize	16,762	24,242	45,720	52,297	52,511	56,441
Millet	5,616	14,830	19,352	17,332	17,851	19,411
Wheat	69	3,763	6,736	6,226	5,341	6,250
Pulses	NA	600	4,288	6,774	7,126	7,793
Total	75,109	1,31,626	2,18,596	2,15,319	2,17,450	2,42,395

Sources: (1) Department of Agriculture, *Arunachal Agriculture in Brief, 1980-93*, Naharlogun, pp. 6-9.
(2) Directorate of Economics & Statistics, *Statistical Abstract of Arunachal Pradesh, 2001*, pp. 2021; 2002, pp. 23-24.
(3) Department of Agriculture, *Annual Operating Plan, 2003-04*, Naharlogun, p. 1.

The foodgrain production in the State during 1970-71 was 75,109 MT which has increased to 2,42,395 MT during 2002-03. But the State is yet to attain self-sufficiency. It may be noted that the coarse-cereals occupy 31.3 per cent of foodgrains and a large part of it is used for preparation of local bear popularly known as "Apong". It is to be pointed out that the cropping pattern is concentrated around rice and coarse cereals as 94.2 per cent of foodgrains is shared by these crops.

It is stated earlier that 62.29 per cent of workers depend on agriculture for their livelihood. Now let us find the share of agriculture in State Domestic Product of Arunachal Pradesh.

Contribution of Agriculture in NSDP of Arunachal Pradesh

The NSDP means Net State Domestic Product which is net value added to the goods and services produced in a region or State or country. Commonly it is known as State income. The share of agriculture proper in NSDP of Arunachal Pradesh is shown in Table 25.5.

Table 25.5: Contribution of Agriculture in NSDP of Arunachal Pradesh

(*At constant prices*)

State/Country	1970-71	1980-81	1990-91	2000-01	2001-02
Arunachal Pradesh	38.4	36.9	34.7	31.2	28.9
All India	44.2	36.3	30.4	23.8	NA

Sources: (1) Bhattacharjee, 2000, *ibid.*, p. 137.
(2) Directorate of Economic and Statistics, Government of A.P., *State Domestic Product of Arunachal Pradesh, 1993-94 to 2001-02*, p. 20.
(3) Department of Economics & Statistics, *Statistical Outline of India, 2002-03*, Tata Consultancy Services Ltd., Department of Economics & Statistics, Mumbai, p. 17.

Here we find a comparative picture of Arunachal Pradesh and all India in respect of contribution of agriculture in NSDP and NDP respectively. The share of agriculture in NSDP of the State was 38.4 per cent during 1970-71 which has decreased to 28.9 per cent during 2001-02 while in all India the decrease is faster from 44.2 per cent during 1970-71 to 23.8 per cent during 2000-01. This shows that the dependency on agriculture is more in this State as compared to all India.

We have seen that over 62 per cent of workers depend on agriculture and about 80 per cent of people live in rural areas and due to this there should be perspective plan in the State for Agricultural and Rural Development.

Constraints and Suggestions

In recent years, there has been serious effort to produce cash crops and commercial crops like tea, coffee and rubber and the value of cash crops was Rs. 164.22 lakhs during 2001-02 as per *Economic Review of Arunachal Pradesh 2002*. The tea gardens were opened in the State during last 2½ decades but this has not changed the agricultural economy of the State and it is concentrated at few selective areas of the State. In order to change the agriculture scenario of the State, the authorities should solve the constraints of the agriculture sector. A few of them are as follows:

1. Lack of road-communication within the State is a prime-constraint. And for that, the topography of the State is the main hurdle. To overcome it, road-communication within the State has to be improved. The dream road of East-West is very essential to tranship the surplus in one district to other districts.
2. Another most important problem of agriculture is marketing problem. Due to non-existence of basic marketing facility the cultivators are discouraged to grow using innovative technique in agriculture sector. The upper and middle belt inhabitants cannot come so easily to the nearby market in Assam. Only the cultivators of the lower belt specially foothills have the chance to come down to plains market.
3. Specific technology for cultivation is lacking in geographically isolated State like Arunachal Pradesh. During rainy season alluvial-top soil drains down to plains of Assam and due to this yield rate is not satisfactory. Thus there must be some technological improvement to contain soil erosion in the State and provide irrigation during the off season.
4. Lack of banking facility is one of the main constraint. Out of 85 CD Blocks in the State only 43 CD Blocks have bank branches. If banking facility and credit to the cultivators are not provided, it will be difficult to improve agriculture in the State.
5. Lack of awareness about the improved varieties, inputs etc. is also one of the main constraints.

6. The mind-set of the present educated youth has to be changed so that they take up the profession of cultivation scientifically specially due to shortage of organised job in the State and for development of agriculture.
7. Fund is one of the main problems in the development of the agriculture. There should be more government fund and investment in the agriculture sector. We have over 62 per cent of workers in the cultivation and only limited amount is spent for them. For example, during the current financial year (2003-04) the government plan fund is revised as Rs. 603.27 crores while agriculture is allotted only Rs. 12.71 crores and horticulture Rs. 11.93 crores which are 2.1 and 2.0 per cent respectively. Besides, the salary component of the employees will absorb a huge chunk of it and very little will be left for true development of the agriculture.

REFERENCES

Bhattacharjee, R.P. (1998). "Agricultural Modernisation in Arunachal Pradesh", in M.C. Behera (ed.). *Agricultural Modernisation in Eastern Himalaya*, Commonwealth Publisher, New Delhi.

Bhattacharjee, R.P. (2000). *Economic Development of Arunachal Pradesh*, Himalayan Publishers, Delhi, p. 36.

Bhattacharjee, R.P., *Insurgency and Development in Arunachal Pradesh*, paper presented in national seminar on 18th November, 2003 sponsored by ICSSR at Itanagar.

Census of India, 2001, Series 13, Arunachal Pradesh Provisional Population Totals, Paper 3 of 2001.

Department of Agriculture, *Agricultural Censuses, 1970-71, 1980-81, 1990-91 and 1995-96*, Government of Arunachal Pradesh, Naharlogun.

Department of Agriculture, *Annual Operating Plan, 2001-02, 2002-03 and 2003-04*.

Department of Agriculture, *Arunachal Agriculture in Brief, 1980-93*, Government of Arunachal Pradesh, Naharlogun, pp. 5-13.

Department of Agriculture, *Arunachal Agriculture at a Glance, 1993-98*, Government of Arunachal Pradesh, Naharlogun, pp. 4-5.

Department of Agriculture, *New Agricultural Policy*, October, 2001, Government of Arunachal Pradesh, Naharlogun.

Department of Economics and Statistics, *Arunachal Pradesh at a Glance, 2002*, Itanagar, p. 4.

Department of Economics and Statistics, *Statistical Outline of India, 2001-02 and 2002-03*, Tata Services Ltd., Mumbai, p. 17.

Department of Horticulture, *Annual Operating Plan, 2003-04*.

Directorate of Agriculture *Agricultural Census, 1995-96*, Government of Arunachal Pradesh, Vol. I, p. 17.

Directorate of Economics and Statistics, Government of Arunachal Pradesh, *Arunachal Pradesh at a Glance, 2002*, Itanagar, p. 4.

Directorate of Economics and Statistics, *Economic Review of Arunachal Pradesh, 2002*, Itanagar, pp. 13-21.

Directorate of Economics and Statistics, *State Domestic Product of Arunachal Pradesh, 1993-94 to 2001-02*, Itanagar, p. 20.

Directorate of Economics and Statistics, *Statistical Abstract of Arunachal Pradesh, 2001*, Itanagar, pp. 20-30.

Directorate of Economics and Statistics, *Statistical Abstract of Arunachal Pradesh, 2002*, Itanagar, pp. 22-24.

Department of Planning, Government of Arunachal Pradesh, *Draft Tenth Five Year Plan and Annual Plan, 2002-03*, Itanagar, p. 17.

26

PRODUCTION OF AGRICULTURAL CROPS AND FOOD SECURITY IN MANIPUR
An Analysis

— RAJMANI SINGH

Introduction

Manipur is situated in the north-eastern corner of India bordering with Myanmar on the east, Nagaland on the north, Mizoram on the south and Assam on the west. The State has a salubrious climate and mountainous topographical structure that provides the State its natural scenic beauty, but on the other hand, it poses problems for economic development for a pretty long time. It is almost rectangular in shape with a precious little valley at the centre, encircled by mountain ranges on all sides. This valley, which is also the "rice bowl" of the State has an area of about 1,851 sq kms. Consisting only about 8.2 per cent of the total area of the State but 64.12 per cent of the total population is concentrated on it. As a result the valley is one of the most thickly populated area of the country.

Growth of Population

The population of Manipur according to 2001 census is 23,88,634 persons (provisional) of which two-thirds of the total population are concentrated in the valley which constitute only one-tenth of the total area of the State. The growing rate of State population is an alarming rate. Table 26.1 shows the comparison of decennial growth rate of population between State and the country. The decennial growth rate of population of the State is much higher than the all India average growth rate. Except 1941-1951 decennial growth rate, the State decennial growth rate exceed the country's average decennial growth rate.

Table 26.1: Decennial Growth Rate of Population

Decades	Decennial growth rate percentage	
	Manipur	India
1941-51	12.90	13.31
1951-61	35.04	21.51
1961-71	37.53	24.80
1971-81	32.46	25.00
1981-91	29.29	23.96
1991-2001	30.02	21.34

Sources: 1. Census of India 1991 and 2001.
2. Press Release, Director, Census Operations, Manipur.
3. *The Telegraph*, Calcutta, March 27, 2001, p. 1.

Production of Crops

The staple food of Manipur includes paddy, maize and wheat, but the wheat crop is grown only in very small pockets and the production of paddy crop is expressed in terms of clean rice after making necessary process.

Although paddy was grown in large scale in all the districts of the State, the largest area under the crop was found to be in Imphal west district followed by Imphal east district which accounted for 15.92 per cent and 14.78 per cent respectively of the total area and the lowest area was in Ukhrul district which accounted only 4.99 per cent of the total area under the crop during the year 2001-02. The estimated area under paddy crop in Manipur was 207.73 thousand hectares for the agricultural year 2001-02.

The estimated production of clean rice for the State was 439.00 thousand tonnes during the agricultural year 2001-02 as against 477.211 thousand tonnes during the agricultural year 2000-01. The per capita production of rice during 2001-02 was worked out to be 186 kgs as against the estimated requirement of 219 kgs per capita per annum. Hence there is shortage of 33 kgs per capita per annum.

The estimated area under maize crop of the State was 16.50 thousand hectares during the agricultural year 2000-01 out of which nearly 48.42 per cent of area are in Churachandpur and Senapati districts. The estimated production of maize for the year 2000-01 was 33.16 thousand tonnes against 24.88 tonnes during the year 1999-2000. Other food crops grown in the State are wheat and pulses. The area and production of rice and maize crops of the State are shown in Table 26.2.

Table 26.2: Area and Production of Rice and Maize from 1980-81 to 2001-02 of Manipur State

Year	Rice		Maize	
	Area (in '000 hec.)	Production (in '000 tonnes)	Area (in '000 hec.)	Production (in '000 tonnes)
1980-81	188.55	273.03	9.91	17.78
1985-86	164.60	332.53	5.74	14.63
1990-91	256.00	452.00	15.50	31.00
1995-96	205.00	420.00	17.00	35.00
1996-97	209.60	420.00	17.00	34.00
1997-98	208.72	381.96	16.00	28.80
1998-99	211.50	397.66	16.50	29.60
1999-2000	126.49	353.99	13.84	24.88
2000-01	212.40	477.27	16.50	33.16
2001-02	207.73	439.00	17.00	26.53

Sources: 1. Directorate of Economics and Statistics.
2. Annual Administration Report 2001-02, Department of Agriculture, Government of Manipur.

If we look at the productivity of rice and maize for the last four years, there are ups and downs in the total production of rice and maize in the State. It was 397.66 thousand tonnes of rice during 1998-99 which declined to 353.99 thousand tonnes in 1999-2000. Again it increased to 477.27 thousand tonnes during 2000-01 but declined to 439.00 thousand tonnes during 2001-02. The main reasons behind the increase and decrease of production in rice in the State are lack of proper irrigation and other infrastructural facilities required for agricultural activities. More than 90 per cent of the total agricultural production is dependent on natural rainfall which is very irregular in nature. Further, to know more details about the target and achievement of crop production during the year 2001-02. The following figures are highlighted below.

Crop Production during 2001-02

Table 26.3 shows the current position of agricultural production target and its achievement for the year 2001-02. The achievement of paddy was 13,723 hectares against the targeted area of 30,000 hectares. The shortfall in the achievement of target was due to inadequate pre-monsoon rain and lack of irrigation facilities.

Table 26.3: Physical Target and Achievement of Area and Production of Crops during 2001-2002

('000 tonnes/hectare)

Sector/Sub-sector		Area		Production	
		Target	Achievement	Target	Achievement
Rice	(a) Ist crop of paddy	30.00	15.72	75.00	37.73
	(b) Local/Improved Paddy	21.28	27.85	42.56	67.86
	(c) HYV Paddy	80.00	71.64	252.86	254.50
	(d) *Jhum* paddy	83.72	82.55	125.56	58.66
	(e) Terrace/wet paddy	10.00	10.00	20.00	20.25
	Total	225.00	207.74	516.00	439.00
Maize	Total	17.30	17.00	34.50	26.53
Wheat	Total	1.75	1.75	4.50	2.00
Potato	Total	8.50	7.00	70.00	60.00
Sugarcane	Total	5.00	3.50	775.00	172.66
Pulses	(a) *Kharif*	4.50	4.00	5.00	3.19
	(b) *Rabi*	21.50	18.00	15.00	15.00
	Total	26.00	22.00	20.00	18.19
Oilseed	(a) *Kharif*	9.35	3.44	10.00	2.92
	(b) *Rabi*	23.00	19.50	15.00	12.00
	Total	32.35	22.94	25.00	14.92

Source: Annual Plan 2002-03 of Agriculture Report, Annexure-II.

On the other hand, due to heavy rainfall during the first week of December 2001, land preparation could not be done upto the desired extent. Consequently, the achievement in the *Rabi* crop coverage was reduced to 15 per cent of the targeted area. Table 26.3 shows the physical target and achievement in terms of area and production in 2001-02. During the year except *Rabi* pulses all other crops were below the target production. The total production of rice is also below the targeted production, i.e., 439.00 thousand tonnes against the 516.00 thousand tonnes. Among the varieties of paddy some varieties such as local/improved paddy, HYV paddy, Terrace/Wet paddy also exceed the target production.

Constraints in Raising Production and Productivity of Crops in Manipur

The following are the main constraints faced by the State while producing different varieties of agricultural crops in Manipur:

(i) Inadequate allocation of fund for the State Plan Schemes in the State Budget.

(ii) Shortage of State Matching Shares, for centrally sponsored schemes during 1997-98. The State's Matching Share in the budget was short of the actual requirement by about 47.29 lakhs restricting the regular flow of central assistance. Even during the year 1998-99 the requirement falls short of the actual requirement by about Rs. 103.03 lakhs.

(iii) Since there are no sufficient certified seed production units/ State seed farms/cooperatives etc. to meet the requirement of certified seeds, difficulty arises in arranging timely availability of such seeds.

(iv) The State is a land-locked area where transportation of practically all agricultural inputs is done from outside the State by road only and as the State is quite prone to land slides/road blockade, timely supply of inputs could not be ensured. Further, lack of buffer stock facilities particularly for bulky inputs like fertiliser, still aggravates the situation.

(v) Lack of infrastructures like irrigation, storage and marketing facilities and natural disasters like flood, droughts in the State stall achievements which often fall short of targets.

(vi) Poor investment capacity of the farmers coupled with lack of credit facilities acts as a spanner in the strategy for increasing agricultural production.

Food Security in Manipur

The valley is one-tenth of the total geographical area with two-thirds of the total population contributing lion's share to the State's agriculture production. Agriculture and allied services account for about 50 per cent of the total net domestic product of the State. Out of the total population nearly 40 per cent of the population are living below poverty line. The crisis of food increases every year with the growing population.

The decennial growth rate of population of the State is very high as compared to all India average as shown in Table 26.1. The decennial growth rate percentage during 1981-91 was 29.29 against the all India average growth rate percentage of 23.86. Again the growth rate during 1991-2001 increased to 10.02 per cent against all India average growth rate of 21.34 per cent. It is clearly shown that the population growth in the State is at an alarming stage. On the other hand cropped area of rice of the State has decreased by 48.27 ha from 256.00 ha during 1990-91 to 207.73 ha during 2001-02. As a result the production of rice has decreased to 13 thousand tonnes from 452 thousand tonnes during 1990-91 to 439.00 thousand tonnes during 2001-02 which is shown in Table 26.2. Above all, the main source of water for agriculture in the State is the seasonal rainfall because the irrigation system is not fully developed. Rainfall is, therefore, the predominant factor for the success or failure of the State crops. Long spells of drought, stormy winds and excessive rainfall exert adverse effect on agriculture. Further, Public Distribution System of the State is also not effectively functioning. As a result people living below poverty line are not benefited as per expectation.

INDEX